HIGH-SPEED LIQUID
CHROMATOGRAPHY

CHROMATOGRAPHIC SCIENCE

A Series of Monographs

VOLUME 1 : Dynamics of Chromatography (in three parts), *J. Calvin Giddings*

VOLUME 2 : Gas Chromatographic Analysis of Drugs and Pesticides, *Benjamin J. Gudzinowicz*

VOLUME 3 : Principles of Adsorption Chromatography; The Separation of Nonionic Organic Compounds, *Lloyd R. Snyder*

VOLUME 4 : Multicomponent Chromatography; Theory of Interference, *Friedrich Helfferich and Gerhard Klein*

VOLUME 5 : Quantitative Analysis by Gas Chromatography, *Josef Novák*

VOLUME 6 : High-Speed Liquid Chromatography, *Peter M. Rajcsanyi and Elisabeth Rajcsanyi*

OTHER VOLUMES IN PREPARATION

HIGH-SPEED LIQUID CHROMATOGRAPHY

PETER M. RAJCSANYI
CENTRAL RESEARCH INSTITUTE FOR CHEMISTRY
OF THE HUNGARIAN ACADEMY OF SCIENCES
BUDAPEST, HUNGARY

ELISABETH RAJCSANYI
SEMMELWEIS MEDICAL UNIVERSITY
II. DEPARTMENT OF GYNECOLOGY
BUDAPEST, HUNGARY

MARCEL DEKKER, INC. New York and Basel

MARCEL DEKKER, INC.

270 Madison Avenue, New York, New York 10016

LIBRARY OF CONGRESS CATALOG CARD NUMBER: 75-29922

ISBN: 0-8247-6325-4

Current printing (last digit):
10 9 8 7 6 5 4 3 2 1

PRINTED IN THE UNITED STATES OF AMERICA

CONTENTS

PREFACE

INTRODUCTION

1. THEORETICAL BASIS FOR HSLC 1
   - 1.1 Thermodynamics of LLC 2
   - 1.2 Thermodynamics of LSC 6
   - 1.3 Kinetic Conditions in HSLC 8
   - 1.4 Factors Affecting Efficiency in HSLC 11
   - 1.5 Optimization of HSLC 18

2. INSTRUMENTATION OF HSLC 23
   - 2.1 Solvent Reservoirs 23
   - 2.2 Gradient Apparatus 24
   - 2.3 Pumping Systems 28
   - 2.4 Sample Injector Systems 29
   - 2.5 Column 30
   - 2.6 Detector Systems in HSLC 43
   - 2.7 Preparative Technique in HSLC 57
   - 2.8 Automation, Computation 60

3. APPLICATIONS 63

3.1 Alcohols, Aldehydes, and Acids 63

3.2 Alkaloids 68

3.3 Amines, Amino Acids, and Azo Compounds 71

3.4 Aromatic Hydrocarbons and Substituted Polynuclear Aromatics 76

3.5 Carbohydrates 82

3.6 Compounds in Biological Fluids and Extracts 84

3.7 Drugs and Related Compounds 91

3.8 Food Constituents 97

3.9 Metal Ions, Metallic Compounds 101

3.10 Nucleic Acid Constituents 102

3.11 Pesticides 106

3.12 Phenols and Related Compounds 110

3.13 Polymer Resins 111

3.14 Steroids 113

3.15 Vitamins 118

3.16 Miscellaneous 121

References 125

List of Symbols 169

Author Index 171

Subject Index 189

PREFACE

Recent advances in liquid chromatography indicate that this old technique is about to enter a new era. Research work carried out during the past few years has provided a wealth of information on the principles, instrumentation, and application of high-speed liquid chromatography, so that a summary may be of value to the liquid chromatographer in most branches of chemistry, and in the fields of medicine, ecology, and pollution. The aim of this monograph is to summarize the state of the art of liquid chromatography, and in doing so we hope to provide investigators in this field with a useful tool.

Although other terms, such as high-pressure, modern, or high-performance liquid chromatography have been used by some investigators, we have chosen the title "High-Speed Liquid Chromatography" because high speed is a functional requirement, whereas high pressure, high sensitivity, and high performance only concern instrumentation.

Our own experiments as well as literature data indicate that the current status of high-speed liquid chromatography has reached a level comparable to that of gas chromatography in the early 1960s, prior to the spectacular advances in this field.

Therefore, we predict a similar dramatic increase in the literature on high-speed liquid chromatography over the next two or three years, during which time the number of papers dealing with high-speed liquid chromatography could increase sixfold.

We wish to thank all our colleagues who have contributed to the preparation of this book, and also Dr. C. Horvath for his continuous interest and valuable help in our work.

Peter M. Rajcsanyi
Elisabeth Rajcsanyi

INTRODUCTION

Although classical column liquid chromatography has been an effective separation method since the beginning of this century, it is still characterized by low column efficiencies and long separation times [1]. The linear velocity of the mobile phase in the column does not exceed 10^{-2} to 10^{-3} cm/sec. With respect to the time required for chromatographic separation, gas chromatography (GC) had certain advantages over liquid chromatography (LC), at least in its early phase of development; the GC method, however, is limited to volatile compounds. Moreover, thin-layer and paper chromatography, despite some advantages, exhibit other limitations. In view of this, an attempt to improve separation by LC seemed to be justified. It was obvious that improving the LC technique to a point where its efficiency and separation time could compete successfully with those of GC would permit application of LC to the analysis of a wide range of compounds. High-speed liquid chromatography (HSLC), including liquid-liquid (LLC) chromatography, gel-permeation (GPC), ion-exchange (IEC), and liquid-solid (LSC) chromatography, was expected to enable chromatographic separation of compounds with molecular weights ranging from 2×10^2 to 10^8.

In this book, the discussion of HSLC will be centered on LL partition chromatography and LS adsorption chromatography. The theory of GPC and IEC will not be treated [2-4]. We shall place emphasis on the following three major steps of development: an appropriate theoretical basis, a satisfactory instrumentation system, and experience in application of the system, all of which are important for reducing the analysis time and improving separation efficiency.

Chapter 1

THEORETICAL BASIS FOR HSLC

The basis of chromatographic separation is the distribution (or partition) of sample components between two phases which are immiscible. Thus separation depends on both the mobile or moving phase and the stationary phase. Interactions between the molecules of the two phases are negligible in GC, but play a very important role in LC. These interactions of the liquid chromatographic system determine the degree of sorption of particular substances and also the effectiveness or selectivity of the separations. (See Fig. 1.1.) As is well known from the general theory of chromatography the retention and dispersion of a solute are determined by both thermodynamic and kinetic parameters [5-8]. It is very advantageous that these two groups of parameters are almost independent of each other and may be separately optimized. On the other hand, it requires full knowledge of the factors affecting peak separation, which are, as follows from the foregoing, rather numerous. About ten years ago the theory developed for GC was believed to be easily applied to LC [9-12]. Although this was realized as early as 1960 by Hamilton et al. [13], by Karr et al. [14] in 1963, and confirmed by Piel [15] in 1966, the rules for high-speed liquid

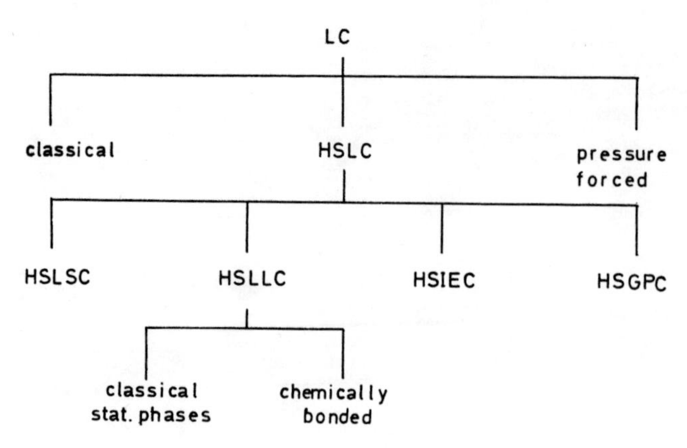

FIG. 1.1 Branches of liquid chromatography.

chromatographic separation are still much less known than for GC. The explanation of this fact lies in the complexity of HSLC.

1.1 THERMODYNAMICS OF LLC

The sample components in a column are continuously partitioned between the mobile and stationary phases in constant ratios, named distribution or partition coefficient [16,17]:

$$K = \frac{\text{solute concentration in stationary phase}}{\text{solute concentration in mobile phase}} \qquad (1)$$

The solute distribution coefficient must be constant during the migration of the solute down the column, that is, a linear partition isotherm is preferred. (At this point several assumptions must be taken into consideration according to Locke and Martire [16].)

For LLC, linear isotherms may be expected up to about 1% volume fraction; for LSC, linear isotherms are rare and the problem is more difficult. If V_s is the volume of stationary phase in the column and V_m is the interstitial volume of the column, the retention volume of a component, for which K differs from zero, is defined as

$$V_R = V_m + KV_s \quad (2)$$

Considering solute residence times due to solute-liquid phase interactions, the net retention volume of a component can be defined:

$$V_N = V_R - V_m = KV_s \quad (3)$$

and from (3) we obtain an equation for the specific retention volume

$$V_g = \frac{V_N}{w_s} = \frac{K}{\rho_s} \quad (4)$$

where w_s is the weight of stationary phase in the column and ρ_s its density.

A new important parameter, the partition ratio or capacity factor, can be obtained by multiplying the partition coefficient by the phase ratio $\beta = V_s/V_m$, i.e., the volume ratio of the two phases

$$k = K\beta = K\frac{V_s}{V_m} = \frac{V_N}{V_m} \quad (5)$$

The partition ratio has theoretically a simple physicochemical meaning: it is the number of component molecules in the moving phase per component molecules in the fixed phase. It means practically the time an average solute molecule spends in the mobile phase relative to that spent in the stationary phase, and can be determined directly from the chromatogram. Taking the condition of equilibrium in LLC at pressure P and temperature T

$$\mu^{s}(T,P) = \mu^{m}(T,P) \tag{6}$$

where μ is the solute chemical potential, s and m indicate the stationary and mobile phases, respectively, we have for the specific retention volume

$$\ln V_g = \ln \frac{\gamma^{m,\infty}(T,P = 1)M_m}{\gamma^{s,\infty}(T,P = 1)M_s\rho_m(T)} + \frac{\bar{P} - 1}{RT}(v^m - v^s) \tag{7}$$

where $\gamma^{m,\infty}$ and $\gamma^{s,\infty}$ are the solute activity coefficients at infinite dilution, M_m and M_s are the molecular weights of the phases and $\rho_m(T)$ is the mobile phase density at temperature T, $\bar{P}$ is the mean column pressure, v^m and v^s are the solute molar volumes in the two phases. This equation allows us to predict partition properties, partition coefficients and retention volumes a priori for systems in which the activity coefficients are known [18-20]. In the dynamic determination of partition coefficients, developed by Huber and co-workers [19,21], a linear relationship is assumed between the retention time and K:

$$t_R = t_0(1 + \beta K) \tag{8}$$

where t_0 is the average residence time of the mobile phase. The unknown partition coefficient of a compound can be calculated by means of Eq. (8) from its retention time and the retention times of other compounds with known partition coefficients measured under the same conditions.

The main sources of error of this method may be adsorption effects and nonlinearity of the distribution isotherm. Partition coefficients of 26 steroids, 6 alkyl benzenes, 5 nitro derivatives of benzene, and 13 pesticides were determined by this method [21]. The apparent limiting activity coefficients were determined by measuring the specific retention volume values of naphthalene, paraffins, olefins, and aromatic hydrocarbons in the range of C_4 to C_{14} in an LLC system, where acetonitrile served as the mobile phase and squalane as the stationary phase [22]. From these values, the excess partial molar free energies, enthalpies, and entropies of mixing were calculated. In the case of two components (1 and 2) in a solute the relative retention

$$\alpha = \left(\frac{\gamma^{m,\infty}}{\gamma^{s,\infty}}\right)_1 \left(\frac{\gamma^{s,\infty}}{\gamma^{m,\infty}}\right)_2 \qquad (9)$$

will depend only on the differences in the solution behavior of the components in the two phases. That is, the relative retention is really a measure of the thermodynamic differences of component distribution, or the difference in free energies of distribution for two components. In LLC one can, therefore, improve separation by changing mainly the nature of the mobile phase; not only that of the stationary phase as in GC. (The improvement of separation in LLC can also be carried out by changing the temperature.) Thus, the use of gradient elution constitutes a bridge between theory and practice.

All in all, from the thermodynamic basis of LLC it is clear that resolution of two adjacent bands can be improved by changing the separation factor or relative retention (α) and the partition ratio (k) [23-27]. In HSLC the role of the capacity factor in resolution cannot be neglected, since the low volume ratio of stationary to mobile phase involves a low k value. These low k values are indeed desired for high speed, as discussed later, but for resolution high k values are preferable. Therefore a compromise must obviously be made, that is, there is an optimal capacity factor value, but with a single solvent k is never optimum for all components of the sample. Thermodynamically, the value of the capacity factor depends on the solvent strength, and strong solvents give smaller k values. Horvath and Lipsky

[28] noted that the peak capacity can be significantly improved by use of gradient elution. In addition to this, we should note that the factors--capacity ratio and relative retention--affecting separation are not independent of each other to the effect that, although k was optimized for every pair of adjacent bands, either a favorable change in the separation factor, i.e., selectivity, or a further increase in column efficiency may be necessary. Alteration in the α value, while keeping k approximately constant, can be achieved by varying the solvent composition. For this purpose mainly practical approaches are at our disposal. Therefore the choice of stationary and mobile phases is one of the weakest points in the development of HSLC [29].

Since the partition coefficient is an equilibrium constant thermodynamically involving Gibbs' chemical potential and activity coefficients, it is dependent on the identity of the solute, the mobile and stationary phases and interactions, as well as upon the temperature [30]. Therefore, both athermal (size) and thermal (energy) factors contribute to the relative retention, i.e., selectivity [31]. Luckhurst, Martire, and Locke developed the above new approach, in order to consider the thermodynamic basis of selectivity in LLC [31,32]. (Their theory is inapplicable to systems where solvent-solvent interaction is stronger than interactions between solute and solvent.) Eon et al. [33] showed that the interfacial tension between the mobile and stationary phases reflects the partition properties of the two phases, and this could be taken as a criterion of choice of the systems to be used in HSLC.

1.2 THERMODYNAMICS OF LSC

In LSC the distribution coefficient of a component between mobile and stationary phases depends on interaction forces; mainly on dispersion forces (nonpolar) and on hydrogen bonding (polar) between the sample component, solvent, and adsorbent. Selectivity of LSC is dependent on the balance established between the two types of adsorption interaction, because the distribution coefficient is determined by the sum of the polar and nonpolar effects. In LSC there is a competition between solvent and sample molecules for the fixed "active" sites on the surface [25,34]. The adsorption of sample molecules requires the desorption of solvent molecules to permit the accommodation of sample molecules on the surface. The capacity factor k is, therefore, determined by the net energy of adsorption of sample molecules, and the net energy of adsorption is given as the sum of interaction energies. Thus, in addition to the chemical nature of the adsorbent surface, the competing interaction should also be taken into account [35-37]. In LSC, as reported by Snyder [38], retention volumes on nonspecific adsorbents are practically independent of the nature of the functional groups, whereas in separations on specific adsorbents they sharply increase when the sample molecules contain polar groups or π-bonds in the case where the solvent cannot undergo specific interaction with the adsorbent. In studying the effect of the composition of binary (cyclohexane + polar) developing solvents in TLC on R_M values of hydroxy-substituted phenols, using silica gel, linear relationships were obtained between R_M values and the logarithmic values of the polar solvent mole fraction, the slope of the lines being related to the molecular mechanism of adsorption [39].

Funasaka and co-workers [40-42] studied the adsorption mechanisms of aromatic compounds on anion exchange resins and the effects of solvents on these mechanisms. They used, among others, Zipax SAX resin for the separation of benzene and naphthalene derivatives. Their results suggest that the adsorption mechanisms are H-bonding or π-bonding between the aromatic compounds and a counterion of the resin, while van der Waals' forces have only little effect on distribution coefficients.

The differences in sorption between strong and weak anion exchangers are generally low, except when using alcohol mobile phases. The electronegativities of the resin counterions do not regularly influence adsorption.

Summarizing the present knowledge on LSC, we may state that from a practical point of view, the selection of the proper solvent strength is more simple in LSC than in LLC, which is a great advantage over the former chromatographic method in HSLC.

1.3 KINETIC CONDITIONS IN HSLC

Assuming that the distribution isotherm is linear and N_{req} plates are necessary to obtain sufficient resolution, the time of analysis t can be expressed as

$$t = N_{req}(1 + k)\frac{H}{u} \tag{10}$$

where H is the plate height and u is the mobile phase velocity [43,44]. The term H/u is a fundamental measure of the speed of analysis. Considering the pressure drop needed to obtain the required plate number, another important parameter can be defined:

$$\psi = \frac{\Delta P_{req}}{N_{req}\eta} \tag{11}$$

where η is the viscosity of the eluent [43,45,46]. If plots of H/u and ψ as a function of velocity are shown as in Fig. 1.2, the kinetic conditions of HSLC can be well understood. The H/u curve decreases very fast at lower velocities and flattens off at higher velocities. The ψ-u curve shows an inverse effect. So there is an optimum velocity range about 0.8-5.0 cm/sec in practice. Taking into consideration the general equation on band-broadening processes, with the assistance of Eqs. (10) and

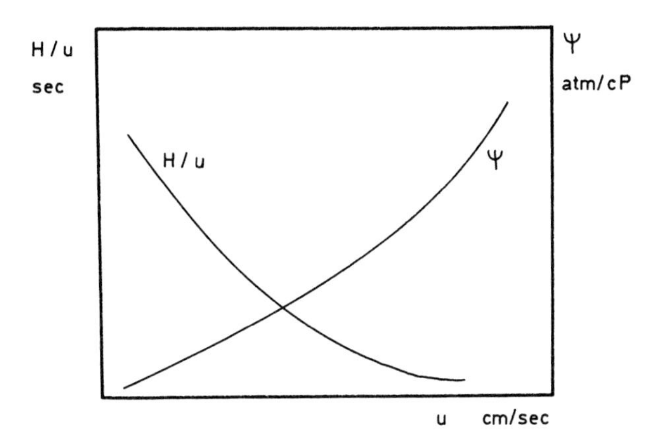

FIG. 1.2 Dependence of H/u and ψ on flow velocity.

(11), overall conditions for high-speed or high-efficiency LC may be derived [5,9,47,48].

The pressure drop ΔP is taken according to operating conditions. The maximum value of ΔP/N is determined by the number of plates needed to achieve the required separation. A decrease of the particle diameter (ΔP/N is proportional to $1/d_p^2$) results in a more than proportional decrease of H but conventional packing materials have some disadvantage for use in HSLC. The fluid velocity is practically constant along the column because of the low compressibility of the mobile phase. In HSLC, because of the coupling of the eddy diffusion and mobile phase mass transfer terms, H/u continues to decrease with the increasing velocity. As pressure drop across the column is related to velocity through the column permeability, the more permeable the column, the shorter the analysis time. Or if analysis time t is maintained constant, a more permeable column permits the use of a longer column, but the length of the column applicable is limited by pressure effects [49-52]. Considering the relation between permeability and the particle diameter, there is an optimum value for both parameters, as permeability may be increased by increasing the particle diameter; larger particles mean, however, higher H values. This means that, in practice, use of irregular packed columns may increase the permeability by at least 1 order of magnitude, while increasing efficiency [53,54].

Taking into account that, in practice, plate height is affected by two counterdispersion mechanisms, by lateral diffusion and the tortuous nature of the flow itself, the effect of fluid velocity, approximately >0.1 cm/sec, on the plate height may be expressed as

$$H = C_0 u^n \tag{12}$$

where n is between 0.2 and 0.8, as it was found by Snyder [55, 56] and other investigators [57-62]. Experimental results indicate that n depends on the particle diameter; it increases for small particles [61] and with the column diameter [63,64].

In a fixed system where only velocity is varied:

$$\frac{H_1}{H_2} = \left(\frac{u_1}{u_2}\right)^n \tag{13}$$

From a practical point of view, a low value of the exponent n is desirable. Using Eq. (13), Waters et al. [57] derived an equation that shows the relationship between resolution, separation factor, analysis time, and pressure drop in the region where Eq. (13) is valid:

$$\frac{S_2 R_1}{S_1 R_2} = \left(\frac{P_1}{P_2}\right)^{(1-n)/n} \left(\frac{t_1}{t_2}\right)^{(1+n)/n} \tag{14}$$

The maximum resolution for a given pressure is achieved if column operating conditions, i.e., particle size, column length, and mobile phase velocity, are chosen so that the term $h^2\varphi$ should be a minimum, where h is the reduced plate height and φ

is a dimensionless parameter related to the permeability of the column [43,59,65].

From the discussion of the theoretical basis for HSLC it is clear that small-diameter columns (1-3 mm), solid supports of particles with diameter less than 60 μm, large pressure drop (up to 7000 psi), and high flow rate (ca. 0.5 to 60 ml/hr) should be used in this new method of separation. These result in a high mobile phase velocity (0.1 to 10 cm/sec), high efficiency in column, and very rapid separation time [66-69].

1.4 FACTORS AFFECTING EFFICIENCY IN HSLC

Column efficiency in HSLC may be represented by the height equivalent of a theoretical plate (H) values and/or the related terms. H may be considered as being built up from a balance between dispersion and relaxation processes which may be divided into two types: (1) mixing processes, and (2) mass transfer processes [5,9,44,70,71,72]. These processes are claimed to be independent and the H value is the sum of four terms, each responsible for a special process:

$$H = H_1 + H_2 + H_3 + H_4 \tag{15}$$

where H_1 is the contribution of mixing by molecular diffusion; H_2 is the result of mixing by dispersion, due to the nonuniformity of fluid velocity within the interparticle space; H_3 is due to mass transfer in the moving phase; and H_4 arises from the stationary phase mass transfer. (In practice extracolumn processes also contribute to the H-value.) We should note that in many HSLC columns, mobile phase effects are predominant compared to the stationary phase effects.

Detailed mathematical analysis of Eq. (15) allows H to be expressed in terms of linear flow velocity, particle diameter, diffusion coefficients, column diameter, and other parameters. The relationship between H and fluid velocity u in HSLC has been discussed above; the influence of other column parameters on column efficiency has also been studied by several investigators [48,56,73-79]. As both the diffusion time and nonuniformity of flow over the column cross section depend on particle characteristics (nature, particle size, porosity, etc.), this subject has been frequently treated.

Results obtained by Halász and Naefe [80] on peak broadening as a function of particle size suggest that a simple relationship between H and d_p is valid within a velocity range of 0.5-5 cm/sec:

$$H = a\, d_p^b \tag{16}$$

where a and b are constant. The average of 80 values shows that b = 1.81 ± 11%, in the particle size range of 50-200 μm, although b slowly increases with increasing velocity. In the particle size range of 11 to 50 μm, b = 1.50 ± 10% was found as the average of 90 values, although these values increased with the increasing velocity (0.3 - 1.65 cm/sec) from 1.47 to 1.51. In evaluating the efficiency and speed of analysis as a function of d_p, Halász and Naefe suggest preference of the smallest possible particle size to HSLC as long as b is greater than 1, because when b is smaller than 1, and N and t_R are constant, the pressure drop increases with the decreasing particle size, indicating that the optimum particle size has been exceeded. This also emphasizes the problem of packing for small particles. These results are in good agreement with those obtained originally by Snyder [55,56], who, however, used b as a constant (b = 1.4) in the following equation:

$$H = C_0\, d_p^{1.4}\, u^{0.4} \tag{17}$$

where C_0 is constant.

Halász and Naefe's results are also in good agreement with those by Beachell and De Stefano [63], Kirkland [59], and Jardy and Rosset [61,81]. For particles of Zipax controlled surface porosity support over 60 μm, Kirkland [59] found b = 1.8 and its continuous decrease as d_p decreased. The least-square curve fitting of values obtained only by the 53, 63, and 74 μm particles gave a b = 1.6 relationship for benzyl alcohol. No dependence of b on the flow velocity was indicated. Jardy and Rosset [61] have, however, indicated that b decreased as the flow rate increased (in the particle size fractions of 18 to 83 μm) and approached a limiting value close to 1.4; this is the reverse of the indication of Halász and Naefe [80]. The variation of H and permeability (K_1) with particle size has been studied by Hamilton et al. [13], Snyder [56,82], Halász et al. [53,83], Majors [84], and Done and Knox [85]. Hamilton et al. [13] found the theoretically expected second-order dependence of the mass transfer resistance constant (C_m) on particle size. Snyder [56,82], however, reported that C_m and K_1 show a lower order of dependence on d_p than predicted: 0.89 versus 2.00 and 1.4 versus 2.00, respectively. The reason for these apparent discrepancies between experimental results and theory probably arise from the complex nature of the packing in a column. It was also found that efficiency is relatively insensitive to a change in particle size only in narrow mesh fractions of the same average particle diameter. Wide mesh-range particle fractions, with lower permeabilities, however, show a loss in column efficiency at a given pressure.

Halász et al. [83] have given an approximation formula for the estimation of the permeability of liquid chromatographic columns packed with particles between 30 and 200 μm. Using this approximation and also experimental data, Halász and Walkling [53] have drawn H-u and K_1-u curves for regular, irregular, and porous layer bead columns. In all cases, the permeability curve was found flattened with the increasing velocity. Majors [84] investigated the effect of particle diameter on column performance and permeability for six silica gels in the particle diameter range of 5 to 40 μm. Both efficiency and permeability were found to depend on $d_p^{1.8}$. Done and Knox [85] determined reduced plate height values as a function of reduced fluid velocity for Zipax fractions of average particle diameters 29, 39, 49, and 106 μm coated with β,β-oxydipropionitrile (BOP). They found, as predicted by theory, that the plot of the reduced plate height against reduced velocity coincided for all particle size fractions, although this has not generally been observed [10,55]. Permeability values increase with the increasing particle size, and suggest that the flow rate is determined by smaller particles present in a given fraction.

Klein [86] studied the effect of the adsorbent porosity on the column performance in LSC and found that an increase in pore size, corresponding to a decrease of specific surface area from 830 to 340 m^2/g, quintupled efficiency. Kiselev and coworkers [36,87] also illustrated the effect of pore size on separation, and achieved the best results on columns with a macroporous silica gel. Kirkland [88] compared a Zipax support with an average pore diameter of about 1000 Å (surface area 0.65 m^2/g), another Zipax support with an average pore diameter of about 500 Å (surface area 0.33 m^2/g), surface textured glass beads with no internal porosity (surface area 0.043 m^2/g), and a Gas-Chrom P support material. The Zipax support (500 Å) gave the lowest h values. Studying the dependence of efficiency on adsorbent activity, Klein [86] also noted a continual increase in H for silica gels of decreasing activity, i.e., of increasing water content. Others [15,89] observed a maximum efficiency at a given H_2O content (activity) or, like Snyder [56], found little difference in separation efficiency for 4 to 20% H_2O/SiO_2. By changing the water content in the mobile phase, the relative retentions were found to change also on alumina packing material [90]. Alumina of five different activity levels with water contents of 1, 4, 7, 10, and 19%, respectively, were used. The H values obtained for the n-heptane inert peak varied between 0.97 mm on the most polar alumina and 1.1 mm on the alumina with lowest activity.

There is some uncertainty about the effect of column length on column efficiency. It was assumed that H is independent of the column length, and studies [77,80,91] on 25- to 400-cm-long columns filled with 80 to 90 μm silica gel-PEG 200 brush-type chemically bonded material and studies [59] on 25- to 150-cm-long columns filled with Zipax packing material have supported this assumption. Recently, however, Majors [84] made experiments on 15- to 100-cm-long columns packed with 13.2 μm silica gel, and found that H tended to increase with the increasing length, although the number of theoretical plates increased linearly with lengths. A similar behavior has been observed by Rajcsanyi [92], using a small particle bonded phase. Majors [84] also noted that both 15-cm and 100-cm columns of 30 μm silica gel gave equivalent H values. Studies on the role of another column geometry parameter, column diameter, suggest a practical optimum at about 0.15 to 0.3 cm in diameter.

The results of Sie et al. [93], Knox and Parcher [54], and De Stefano and Beachell [63,64,94] provoked more interest. The improvements observed in column performance obtained by using large diameter columns (about 10 mm) may be the result of a phenomenon called "infinite diameter effect." Knox and Parcher [54] have found in their theoretical work that for ideal columns h should be a universal function of reduced velocity but this function may depend upon the column to particle ratio because of wall and transcolumn effects. They suggested that when using a column with a diameter considered "infinite" in which the solute, diffusing laterally as it moves through the column, could never reach the walls before it was eluted, H values could be significantly reduced.

De Stefano and Beachell [63,64,94], using the equation by Knox and Parcher [54]

$$d_c > (2.4\ d_p L)^{0.5} \tag{18}$$

for calculation of the minimum column internal infinite diameter, demonstrated the applicability of this theory in LSC and in HSLLC, too. Sample volumes over 900 μl and sample weights over 0.6 mg caused a sudden increase in the H values and a decrease in resolution. For columns that have large diameters, but not large enough to produce the infinite diameter effect, i.e., columns that are near the borderline of having "infinite diameter," unusually high plate heights can be observed, probably owing to transcolumn inequilibrium [63]. Observation of column packing materials shows that in HSLC all the supports by which mass transfer distances within the adsorbent particles can be reduced have a possibility for wide application. Support materials and column packing procedures will be discussed in detail later. Although reported GC data indicate that application of narrow particle distributions seems desirable in order to prevent a decrease in efficiency, HSLC studies showed permeability and the H versus u curve to be practically independent of the width of the sieve fraction [59,84,85]. The conditions became worse only with very broad (25-150 μm) fractions [80].

In HSLLC, the level of stationary phase loading is one of the most important parameters affecting efficiency. Frolov et al. [95] observed a continuous increase in the curve of H versus the amount of BOP stationary liquid phase with increasing liquid phase amount using silochrom supports. In order to shorten the separation time a decrease in the stationary phase amount is recommended. A similar increase on the curve of H against liquid phase amount was observed by Randau and Schnell [96], using a silica gel adsorbent of different particle sizes and 1,2,3-tris(2-cyanoethoxy)propane (Fractonitril III) stationary liquid phase. The advantage of columns heavily loaded with BOP was discussed by Engelhardt and co-workers [97,98]. They found that the amount of liquid stationary phase must be greater than 0.6 g/g on an optimal silica support with the following characteristics: specific surface area of 350 to 500 m^2/g, pore volume of 1 to 1.3 ml/g, and average pore diameter of 100 to 200 Å. The main advantage of the heavily loaded columns is the potential for large sample sizes. The thickness of stationary liquid films has great importance also in the case of pellicular packing materials. Kirkland [88] studied columns packed with porous layer bead supports having a surface area of 0.65 m^2/g in which the weight of stationary liquid phase was varied between 0.2 and 2.0%. The mass transfer in the stationary phase was limited at liquid loadings above 1%, indicating that the H_1 term in Eq. (15) was beginning to show a significant contribution to H. The average film thickness calculated at the optimum level of 0.5% liquid phase is about 100 Å, meaning approximately 10 monolayers.

Studying the effect of sample size on H, Snyder [56] found that the increase of the sample size up to the linear capacity of the column brought about a slight increase in the H value, and beyond this it showed a sudden change with further increase of sample size which caused a worsening in the separation. At low values of sample weight:adsorbent weight ratio (below 5×10^{-3}), both H and the sample equivalent retention volume, which is the retention volume corrected for column void volume and divided by adsorbent weight, were found to be independent of the sample size [99]. The effects of sample size and concentration on H in infinite diameter columns indicate that a 200-fold increase (from 10 μl to 2.0 ml) in sample volumes causes only a relatively minor loss in column efficiency [64]. Injection of constant solute weights shows that large sample volumes of dilute solutions allow more efficient separations than small concentrated solution volumes. The linear capacity of the infinite diameter column is smaller for concentrated solutions than for diluted solutions. Both α and K are essentially constant for sample sizes below the column linear capacity. Engelhardt and Wiedemann [90], using alumina of different activity, observed that beyond a given sample size the net retention volume decreases with increasing amount of sample. Since isotherm linearity decreases with higher activity, sample size must be the smallest on alumina of highest activity. Although it is evident from Eq. (14) that resolution and efficiency depend on pressure, in practice, continuous increase in pressure is not too advantageous because five- or sixfold increase in pressure gives only about onefold increase in efficiency [45,57,100]. As mentioned earlier, extracolumn effects also contribute to the H value [45,83,101,102]. The influence of the HSLC equipment cannot be neglected, considering extracolumn contributions to H, if the measured plate height in two identical columns of different length is equal. Plotting H values against the flow rate for different ion pairs gives different lines which are not parallel. This indicates α-dependence on ion-exchange reactions [61,62,81]. The influence of wall effects on resolution and efficiency has also been studied, using metal and glass tubes [103,104]. For the transfer of TLC separations to LSC or HSLSC, glass tubes are recommended.

1.5 OPTIMIZATION OF HSLC

It is evident from the foregoing that optimum separation in HSLC requires minimum $h^2\varphi$, as a combination of high efficiency, high selectivity, and appropriate capacity. It is understood, however, that at the present stage, optimization procedures are based mainly on kinetic and not on thermodynamic parameters. In this section, the authors attempt to give a brief summary of optimization procedures, both from kinetic and thermodynamic aspects.

In optimizing an HSLC separation of a given sample mixture at given stationary and mobile phases, one will face a multiparameter problem involving the mutual dependence of plate height, column length, fluid velocity, column pressure, and particle parameters. Regarding these conditions, certain thermodynamic parameters are fixed, such as α, K, viscosity (η), diffusion coefficient, etc. We may then choose between two ways of optimization: (1) maintaining a constant resolution (R) and reducing retention time as much as possible (minimum time analysis) [11,69,105-112], and (2) obtaining a maximum number of effective plates in a fixed time, i.e., keeping t_R constant and making R as large as possible (time normalization) [82,110, 112-119]. By the first method, defining R = 1 as an adequate resolution, we may calculate retention time as follows

$$t_R = \frac{\eta L^2 f_1 (1 + k)}{K_1 \, \Delta P} \tag{19}$$

where f_1 is the total porosity in the column [112]. Among others, Knox and Saleem [110] attempted to derive general equations for some optimized parameters. In a minimum time analysis conception, when the pressure drop is limited, for the analysis time they had

$$t = \frac{S^2 h^2 \varphi \eta (1 + k)^5}{k^4 \, \Delta P} \tag{20}$$

where S is the separation number as defined by Purnell. From Eq. (20) it is clear that for a given S and P, a column should be designed which operates at the minimum value of $h^2\varphi$. Equation (20) is of practical use only when taken in conjunction with the appropriate equations for d_p and L. For HSLC, these equations of optimum separation require a pressure of about 3000 psi, column length of 10 cm, column-particle diameter ratio 1000, i.e., a particle diameter of 2 μm, to achieve a minimum separation time of 90 sec, a separation number of 10^4, and both reduced plate height and fluid velocity of three. These conditions, however, mean also a maximum injection volume of 10^{-3} ml and a maximum detector working volume of 5×10^{-3} ml.

More recently, Majors and MacDonald [111] have attempted to give a general equation for optimum performance, using a minimum time analysis conception:

$$t = \frac{(C_0N)^{2/(1+n)}(\eta K_2)^{(1-n)/(1+n)}(1+k)d_p\{b-[(1-n)/(1+n)](y-b)\}}{\Delta P^{(1-n)/(1+n)}} \tag{21}$$

where $K_2 = d_p^y/K_1$, the column permeability parameter. Experimental data have also been obtained from studies of effects of particle diameter on efficiency, permeability, and shape of H versus u curves for porous silica gel and alumina of particle diameter between 5 and 40 μm [111].

By the second method, time normalization, one may normalize via changes in fluid velocity, column length, and particle size, in relative retention or in a manner to achieve maximum resolution per unit time. From a practical point of view, first one should choose an initial experimental system, and then choose a normalization technique best suited to one's experimental conditions. The effect of normalization should be evaluated on the basis of the general expression for resolution:

$$R = \frac{1}{4} N^{0.5}\left(\frac{\alpha - 1}{\alpha}\right)\left(\frac{k}{1 + k}\right) \tag{22}$$

which contains the efficiency term ($N^{0.5}$), selectivity term

$[(\alpha - 1)/\alpha]$, and capacity term $[k/(1 + k)]$ and where N and k are recommended to refer to the second peak [5,82,113]. Jolley, Chilcote, and co-workers [115-118] applied normalization techniques, using computers in the identification of chromatographic peaks by their elution position to well-defined reference peaks. In these procedures, the use of two reference compounds is more advantageous than that of only one compound [119]. The peak capacity of a chromatographic system may also play an important role in optimizing HSLC separations. Its dependence on several factors has been discussed by Scott and Snyder [120-122]. Recently, considering the different optimization theories, Snyder [66,67] presented a simple, rapid, practical approach to predicting nearly optimal conditions for a satisfactory separation in HSLC.

In summarizing the present knowledge of optimization, it should be emphasized that separation in HSLC is the function of so many parameters that its mathematical optimization involves a great number of problems. It seems that, considering all relevant parameters, mathematically speaking, it is probably impossible to find a general solution to this problem. In addition, one should also keep in mind that mathematical optimization may result in practically unacceptable requirements. On the other hand, we must take into account facts and calculations which indicate a gap of several orders of magnitude between the theoretical and actual performance of present-day's liquid chromatographic columns [123,124]. This requires more effort toward optimum conditions in HSLC separations.

Chapter 2

INSTRUMENTATION OF HSLC

Among the general reviews frequently appearing on HSLC [125-170], survey papers dealing with HSLC equipments [155-170] use mainly a similar block diagram of a typical HSLC system as shown in Fig. 2.1. The parts of such a system will be discussed in the order in which they appear in the diagram.

2.1 SOLVENT RESERVOIRS

Reservoirs of stainless steel or inert polymer containing about 1.0 to 1.5 liters liquid are optimum for analytical purposes [171-172]. For chromatographic systems used in preparative work, reservoirs of larger volume should, however, be selected. Under particular conditions glass can also be used as reservoir material. Accessories in the reservoir are a heater, a (magnetic) stirrer, and inlets for vacuum and nitrogen purge. Attached to the reservoir may be a cooled condenser which helps reduce losses of volatile solvents used as carrier [173]. Degassing the polar mobile phases to prevent the formation of bubbles, which can be troublesome with detectors, can be accomplished by vacuum [174]. Complete removal of dissolved oxygen may require heating and nitrogen purge. The reservoir can also serve as a liquid pump.

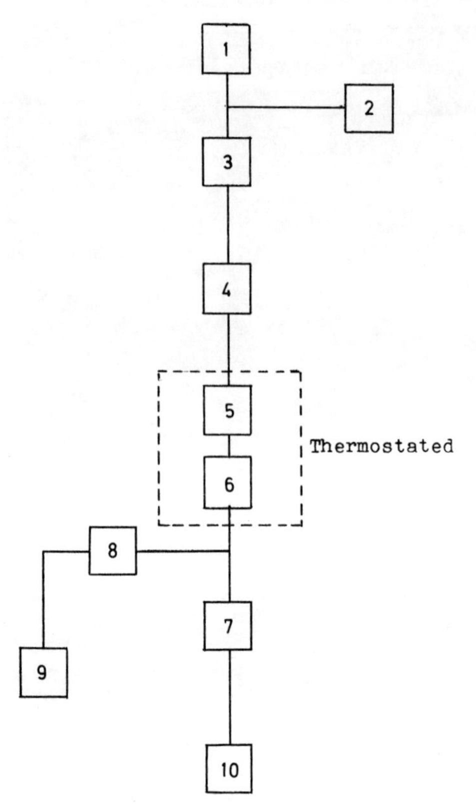

FIG. 2.1 Block diagram of a typical HSLC system. (1) reservoir, (2) gradient elution device, (3) pump, (4) injector, (5) column (precolumn), (6) detector, (7) recorder, (8) flow meter, (9) fraction collector, (10) data system.

2.2 GRADIENT APPARATUS

Gradient elution or solvent programming can be defined as changing the chemical composition of the mobile phase during an analysis [175,176]. In this way the k value of each eluted band is approximately optimum during its movement through the column; therefore, gradient elution is the most useful approach in improving separation. Gradients may be stepwise or continuous.

Gradient elution devices can be divided into two categories: (1) mixed-stream devices, and (2) mixing-chamber devices. Today, most commercial manufacturers offer also gradient elution accessories; the optimum gradient for a particular separation is, however, selected mainly by trial and error. Gradient elution can provide significant improvement in resolution at all points in the chromatogram, particularly at the front end, by adjusting the k values of individual bands so that each band is eluted from the column under near-optimum conditions. The role of gradient elution may be especially important in the analysis of unknown samples or with samples whose components show wide variation in structure and polarity.

Snyder and Saunders [177] discussed the problem of designing an optimum solvent program for a given sample mixture and selected requirements for this program. In this optimization, they have chosen the maximum sample resolution per unit time as primary goal and discussed the requirements of the solvent program separately. The elaborated solvent program was used for the separation of a complex test mixture consisting of 17 substituted benzene and naphthalene compounds as well as heterocyclics, from chlorobenzene (eluted first) to 3,4-benzacridine (eluted last). Scott and Lawrence [178], supported by earlier experiences concerning this problem [179-181], discussed gradient elution under conditions of axial equilibrium, i.e., without polarity gradient, produced an apparatus for carrying out this purpose, and illustrated its applicability. In the system consisting of a precolumn and a column filled with silica gel and connected in series and a heptane/iso-PrOH mixture used as the mobile phase, the increase of the mobile phase polarity is developed by temperature programming the column and precolumn. This temperature programming results in an increase of the iso-PrOH concentration in the mobile phase, occurring uniformly throughout the length of the column. Applicability of this system was illustrated by analyses of a synthetic mixture of squalane, methyl palmitate, dinonyl phthalate, and tristearine, and a high boiling fraction of essential oil. Scott [182] developed theories for gradient elution using the continuous and incremental methods of solvent mixing, including a simple computer program that provides curves relating moving phase composition with time for the incremental method of mixing.

Scott and Kucera [183-185] elaborated a procedure for choosing a series of solvents for incremental gradient elution in LSC which method of solvent programming was achieved by changing the mobile phase supply through a sequence of different solvents in a stepwise manner; and constructed an apparatus for use of incremental method of gradient elution. The solvents, including n-heptane, CCl_4, $CHCl_3$, ethylene dichloride, 2-nitropropane, nitromethane, propylacetate, methyl acetate, acetone, EtOH, MeOH, and water, were used to separate a test mixture commencing with squalane and ending with glucose. The time taken for the separation of the test mixture was 7 hr, involving an actual rate of one k value every 16 min. This system has been broadened by H. Reeve Angel Co. [186] with two series of solvents, one for initial use and one for refining preliminary screenings, which are compatible with UV detection.

Byrne et al. [187] developed a multifunctional gradient elution device which could produce a variety of shapes and time durations without hardware changes. Using this device, liquids from two reservoirs could be mixed automatically by two electronically controlled proportioning valves. This system was applied in the separation of chlorinated biphenyls on a Permaphase ODS column, using linear and exponential gradients. Delfel [188] described an apparatus for linear gradient elution under pressure, which was used in the chromatography of plant sapogenins and steroids on silicic acid columns. Chilcote,

Scott, and co-workers [189,190] developed an automated two-chambered gradient-generating system which could provide also the automated regeneration of the column in preparation for the next analysis.

The refractive index and microadsorption detectors which respond to changes in solvent composition cannot usually be used with continuous gradient elution. At UV detector, since solute UV absorbance is a function of the solvation strength, calibration runs should be made for quantitative works. Using multicomponent eluents to provide the eluent strength necessary for separation, at the above-mentioned detectors, spurious peaks were observed. Solms et al. [191] have set up a model to account for such behavior, used computer simulation, and verified the results of the computer study both quantitatively and qualitatively. They found that in certain cases quantitative errors could amount to 30% of the correct result. However, with proper attention to solvent selection, a refractometer can be used with gradient elution as demonstrated by Bombaugh et al. [192] in the separation of UCON 50HB55 and Triton X45 polymers, using a solvent program of 10 to 70% iso-PrOH in hexane and that of 5 to 80% iso-PrOH in hexane, respectively. In situations where initial and endpoint compositions of the mobile phase can be chosen with similar refractive indices, separations by gradient elution can be carried out [193,194]. This mode of monitoring combined with gradient elution is applicable only with a reduction in sensitivity of 10 or more.

Katz [195] studied some pH control problems in gradient elution, using anion-exchange chromatography; pH of the eluate formed from 0.015 N to 6.0 N acetate buffer was recorded from runs with and without a urine sample. A sharp rise of pH following an abrupt increase in buffer concentration and a sharp decrease of pH following an abrupt drop in buffer concentration was observed. After an initial gradual increase in buffer concentration, a broad peak occurred. Regarding the dimension of these pH changes, it was found that a sudden change in buffer concentration from 0.015 N to 6.0 N caused a rapid increase in eluate pH from the original 4.4 to 5.65.

2.3 PUMPING SYSTEMS

The small particle sizes and small diameter columns being used in HSLC require pumps delivering mobile phase at pressures in the range of 300 to 7000 psi [15,196-201]. Bidlingmeyer et al. [202,203] used a high-speed liquid chromatograph in the separation of methyl- and ethylorange, which delivered eluent under pressures up to 60,000 psi. The advantage of the screw-driven type mechanical pump is that it can deliver a pulse-free supply of liquid, but it has limited solvent capacity. The main drawback of the reciprocating piston and the diaphragm pumps is that these produce a pulsating flow. One of the major objections to the use of a pulsating pump for HSLC is that optimum circumstances are difficult to realize when flow is varying. A specific advantage of the reciprocating pump is that it has small internal volume and its delivery is continuous. Pneumatic pumps are operated by gas pressure working on a container or piston that pressurizes the mobile phase [171,204,205]. The advantage of these pumps is that high pressures can be easily achieved and they deliver a pulse-free supply of liquid phase. By varying the gas supply with a pressure programmer, the mobile phase flow can be easily programmed [206-209]. Karger and Berry [210] presented a continuous gas displacement pumping system that offers no pulsation and inexpensive construction as well as low detector noise level. This pump has been operated over 2000 psi.

When using flow-sensitive detectors such as RI, UV, and microadsorption, etc. in order to achieve the lowest detection limit, nonpulsating pumps are preferred. Gradients may be generated by external or internal mixing methods, using one or two pumps; nevertheless for externally generated gradients, reciprocating pumps are preferred and for internally generated gradients, gas amplifier and syringe-type pumps are recommended. When using the latter type of pumps, two pumps are needed to direct solvent flows to a mixing chamber [211]. Electronic control is advantageous for performing both solvent and flow programming [211,212]. For nonroutine applications where solvent programming may be necessary, a Varian LC-400 HSLC apparatus was modified by a system of valves to allow better changing of the mobile phase and the use of a direct on-column stop-flow injection [213]. Control of flow rate, which is particularly important in flow and solvent programming, has been discussed by Halász and co-workers [208,214]. For HSLC systems using pumps which produce pulsations, pulse dampers are required. In some cases, multihead pumps are applied to eliminate pulsations. Flow-through Bourdon tubes can also be employed as pulse dampener. For more detailed discussion of pumps used in HSLC, the reader is referred to the excellent review of Berry and Karger [215] as well as that of Henry [172].

2.4 SAMPLE INJECTOR SYSTEMS

Two modes of sample injection are used: (a) with moving solvent; and (b) with stopped-flow technique. Very high separation efficiency can be obtained by on-column injection; injecting the sample directly onto the top of the column packing. The major advantages of this type of injection are as follows: simplicity in design, injected volume readily changed and giving only very small contribution to elution bandwidth [198,216-219]. Injections under pressure can be carried out by using syringe-septum, syringe-septumless, and valve-injection systems. A great number of factors may influence the performance of an injector: its design, dead volume, "fountain-effect," type of septa, leakage of the syringe, penetration of the pressurized septum, limited sample volumes, and wasting in valves during the loading step.

Sample valves are most often used in HSLC above 1000 psi, or when high repeatability of sample introduction is required [220-223]. One of the major advantages of sample valves is that they are easily adapted to automatic repetitive analysis [224]. Injection ports are often located at the bottom of the column to decrease the time necessary to sweep all air out of a newly installed column on startup [171,202]. Smuts and co-workers [225] have designed an inlet system based on splitter principles which decreases considerably the solute bandwidth. A sampling system was described by Krejci et al. [226] by means of which compressed samples could be transferred into a pressure-forced (up to 25 atm) liquid chromatograph without previous sample release. Cassidy and Frei [227] designed a simple, inexpensive injection port for stop-flow technique, as well as a pressure relief device. Lindley et al. [228] designed and constructed an automatic sample loader which is controlled by an electronic programmer. The device seems to be applicable, however, only at low pressures. Finally, two reviews dealing with the sample loading technique should be mentioned [215,229].

2.5 COLUMN

The success of an analysis may depend on the column in great measure. The column material can be glass or stainless steel in several forms and lengths [230]. Column materials of precision bore and seamless stainless steel, alumina, and copper gave almost equivalent efficiency results as demonstrated by Karger and Barth [231] separating a mixture of toluene, benzyl alcohol, and phenol on a 1.1% BOP-Corasil column, using heptane as the mobile phase. Teflon-coated alumina columns were not reproducible. The influence of coiling on the efficiency of the separation has been shown to depend on column and coil radii [232-234]. This influence in HSLC is increased as the tube radius is increased or the coil radius decreased [233]. When using open tubes in LC, an increase in efficiency was observed due to an increased lateral mass transfer caused by secondary flow [235,236]. Coiled columns for use in HSGPC were examined by Heitz [237]. Using a column 2 mm in diameter and a coiling radius of 10 cm, the additional contribution to H due to coiling is negligible with oligostyrenes of different molecular weights. For separating polymers, a decrease in column radius seems to be advantageous.

Column lengths can vary from 15 to 300 cm or more (longer columns generate higher back pressures) [105,238,239]. The typical diameters of the analytical columns in HSLC range from 1.5 to 3.0 mm. As the column internal diameter is increased, an increase in H occurs; with very small internal diameter columns, extracolumn peak-broadening effects, however, become significant and it is possible that wall effects are a dominant factor [88,240]. In preparative applications, columns with larger diameter are used. The design of column connection tubes for HSLC is also important, and a careful design can restrict the increase in bandwidth to a maximum of 5% for all eluted peaks [241]. An analytical column of usual dimensions (1 m x 2 mm) contains about 6 to 8 g of a superficially porous packing material and 2 to 4 g of a regular porous packing, depending on particle diameter.

The influence of column packing, coiling, and length on the peak-broadening is codetermined by the surface properties, mechanical stability, and sieve fraction of the support [80]. For LSC, a detailed discussion on the size and geometry of the adsorbents was presented by Snyder [55] as well as by Wohlleben [242]. In HSGPC, semirigid [243-246] or rigid [247-253] packings can be used. Whenever possible, the column in HSLC is dry-packed; for particles below 35 μm and irregular particles below 45 μm, a slurry-packing technique is recommended [46,88, 93,254]. Techniques for dry-packing HSLC columns are not much different from those usually used in GC; i.e., a tapping-rapping technique is favorable, without vibrating. An alternative method may be tamping [79,255].

Scott and Lee [256,257] applied a dynamic packing technique for filling HSIEC columns. In this technique, ion-exchange particles with diameters less than 20 μm are forced into a column in a flowing fluid at a velocity greater than their settling velocity. Due to this packing procedure, there is no size segregation in the column bed. The best result can be obtained when the column is first filled with clear liquid before ion-exchange particles entering the column. First a reservoir of larger diameter than the column to be packed may be filled and then extrude the particles into the small-diameter HSLC column by a liquid.

When using small-particle (5-10 μm) materials, HSLC columns may be packed by a special technique called balanced density slurry technique [60,254,258,259]. In this technique, adsorbents are suspended in an appropriate mixture of balancing solvents, such as tetrabromoethane-tetrachloroethylene or tetrabromoethane-dioxane-carbon tetrachloride. The composition of the solvent mixture must be changed until particles do not settle, but are poised. The balanced slurry obtained in the above manner is placed into a reservoir on which a column is attached. After filling the reservoir with water, the system may be pressurized to a high pressure to pump the slurry into the column. The system should be maintained under pressure until the first drop of water appears at the end of the column. The water remaining in the column should be removed by aspiration or pumping other fluids (depending on the type of the packing) through the column. Comparing dry-packed and balanced density slurry-packed columns, Kirkland [254] found better results in H values for the latter type packing technique. The remarkable review of Kirkland [260] discussing columns for HSLC should be mentioned.

Supports for HSLC

Two main categories of column packing materials in HSLC may be distinguished: porous and superficially porous (pellicular or porous layer) supports [260-262], although the packing materials in HSLLC, HSLSC, HSGPC [263], and HSIEC [240] have different characteristics. Chemically bonded packing materials signify a special type of packing materials and show the main tendency of development as discussed later. Classical porous packing materials are either spherical or irregularly shaped, have deep pores, large surface areas (50-400 m^2/g) and, therefore, high sample capacities. Porous supports include silica gels, alumina, and diatomaceous earth materials.

Pellicular packing materials in HSLLC and HSLSC consist of a solid core and a thin, porous coating, and are commercialized as Zipax, Corasil I and II, Vydac, Perisorb, Pellosil, Pellidon, etc. Pellicular ion-exchange materials for HSIEC were developed by Horvath et al. [264] in 1967, as the first major result in the field of new supports for HSLC. A thin film of styrene/divinylbenzene resin polymerized onto a spherical glass bead was treated to make either a cation or an anion exchanger. Resins have also been coated on Zipax porous layer bead supports [230, 265,266] and other supports [267,268]. Pellicular (superficially porous) resins are of 3-12 μeq/g capacity which requires sensitive detectors, but are of high efficiency. An excellent review on pellicular ion-exchange resins in HSIEC has been recently presented by Horvath [269]. Porous ion-exchange resins previously developed mainly for amino acid analysis can also be used, although high crosslinking (above 8%) is desirable. Small particle size porous packing materials are also used in HSLC [60, 254,258] to diminish the depth of the porous packing pores, which means minimizing diffusion pathlengths during separation.

A list of packing materials most frequently used in HSLC is given in the article of Leicht and De Stefano [262]. Column packings in HSGPC differ from those of the other HSLC modes and are discussed elsewhere [245,246,270-272]. The performance of several different packing materials in HSLC has been evaluated by comparison of through parameters characteristic of the efficiency. Kennedy and Knox [273] evaluated the performance of Porasil, Corasil, Corning Porous Glass (CPG), and Zipax packings for retained and unretained solutes in HSLLC and HSLSC by comparison of plots of log h versus log v. Zipax cannot be used for HSLSC owing to its small adsorptive capacity. The other supports give similar log h versus log v curves for retained solutes (acetophenone and tropanol) in HSLSC, but Corasil shows the best results for an unretained solute (chlorobenzene). In HSLLC, where these supports were coated with BOP, Zipax shows the best performance. No significant difference was found in the performance of Corasil I and Corasil II, although their surface areas are different. The results obtained by Kennedy and Knox [273] are generally in good agreement with those of Little et al. [274] and Vries et al. [275].

Corasil I and II, Perisorb A, and Zipax packings were compared and characterized as to their particle size and pore structure by Unger et al. [276]. These pellicular packings were also tested in HSLLC where Fractonitril III was used as a stationary phase. Their results compare well with those of Kennedy and Knox [273]. In HSLSC, using a test mixture of benzene, m-diphenyl, m-terphenyl, m-quaterphenyl, and m-quinquephenyl and n-hexane as the mobile phase, the Corasil and Perisorb columns showed better selectivity than Zipax. In a comparison between Zipax and porous silica microspheres (5-6 μm and 8-9 μm), the latter packings gave better results [60]. These small particle silica gel packings show retention and efficiency characteristics less dependent on water content in mobile phase than wide-pore silica packings [277]. The performance of chemically bonded packing materials (Permaphase ETH and ODS) was studied by Knox and Vasvari [278]. The data interpreted in terms of the equation $h = B/v + Av^{0.33} + Cv$ indicate relatively slow mass transfer in the stationary phase for Permaphase ETH. When Permaphase ODS is used, the asymmetry of peaks increases with the increase of k. The calculated values of mass transfer heats show that retention on Permaphase ODS is largely determined by the mass transfer process between the stationary and mobile phases, whereas for Permaphase ETH, heat of transfer is not the only factor determining k.

Sizing of porous support materials by dry sieving can be achieved down to 30 to 40 μm [46]. Scott [279] developed a continuous elutration method for preparing spherical ion-exchange beads of 5 to 40 μm by fractionating the material as a suspension into an upward flow of liquid. Irregularly shaped porous silica gel supports are commercially available down to 10 μm and spherical microsilicas down to 5 μm and can be used in HSLC. Unger et al. [280] elaborated a procedure for the preparation and sizing of porous silica microspheres, which allowed obtaining a particle with narrow size range and given pore diameter.

Tesarik [281,282] graded packing materials by flotation in a constant stream of gas or liquid. Pressure and flow-rate regulation are the most important factors in obtaining the best grading, i.e., the maximum sharpness of grading. Majors [84] sized and graded commercial porous silica gel by a centrifugal-air particle classifier. The standard deviation values calculated from frequency distribution curves were between 1.26 and 1.46.

Particle size distributions for Zipax fractions are presented by Done and Knox [85]. Pore size distribution and total pore volume of pellicular packing materials can also be examined by the use of narrow molecular weight distribution polystyrenes and HSGPC [283]. The adsorbent materials recommended for HSLSC can be coated with stationary phase and used in HSLLC.

Chemically Bonded and Surface-Modified Supports

The conventional packing materials to be used in HSLC have procedural limitations: the carrier must be presaturated with the stationary phase to prevent its stripping from the support [128]; high flow rates are to be avoided; gradient elution technique cannot be used. To surmount these disadvantages, chemically bonded packing materials have been developed [284-290]. Halász et al. [291] developed the so-called "brushes," esterified siliceous supports which were commercialized as Durapak. The brush packings, however, tend to be thermally and hydrolytically unstable; water and alcohols limit their general usefulness.

A second type of bonded phases is prepared by polymerization after silylation of the hydroxyl groups on the silica surface as can be seen in Fig. 2.2. This type of packing is thermally and hydrolytically stable [254,292-294]. Locke et al. [295] have given procedures for forming silicon-carbon bond being thermally and solvolytically stable. The primarily

```
                                                    / Me
Si          Si - OH        Si - O - Si - Cl
|\           |                |           \ Me    polymerization
| \   H2O    |      DMCS      |
O  O  ---->  O     ------->   O                   ------------->
| /   H+     |                |            / Me
|/           |                |
Si          Si - OH        Si - O - Si - Cl
                                          \ Me


          Me      [ R         ]       / Me
          |       [ |         ]
Si - O - Si -    -[ Si - O -  ]  - Si - Me
|         |       [ |         ]       \ Me
|         Me      [ R         ]n
O
|         Me      [ R         ]       / Me
|         |       [ |         ]
Si - O - Si -    -[ Si - O -  ]  - Si - Me
          |       [ |         ]       \ Me
          Me      [ R         ]n
```

FIG. 2.2 Preparation of a chemically bonded phase.

substituted supports were also transformed into their ion-exchange forms. Reverse-phase column packing materials have been recently developed and commercialized as Bondapak phases [296-299]. Brust et al. [300] prepared ≡Si-N= bonded phases with one or more functional groups such as -COOH, $-SO_3H$, $-NH_2$, -CN, $-NO_2$, $-C_{18}H_{37}$, -CHPh, and $-CHC_5H_4N$. Relative retention times of benzene, phenanthrene, p-nitrotoluene, p,p′-DDT versus anthracene were also determined. Unger and co-workers [301-305] prepared, in two consecutive reactions, bonded phases of the type Si-R (where R was different aryl and alkyl groups) for GC. Hill [306] used silica gel as an insoluble support for the preparation of 8-hydroxyquinoline bonded phase which could be used for metal chelation. Silica gel was treated with 3-amino-propyltriethoxysilane and after benzoylation, reduction, and diazotization coupled with 8-hydroxyquinoline.

Novotny et al. [307,308] managed to prepare polar-bonded phases, with hydroxy, cyano, and ester functional groups by polymerization of trichloro[3-(4-chloromethylphenyl)propyl]- and dichloromethyl[3-(4-chloromethylphenyl)butyl]silanes and subsequent modification. The values of k and N/t_R as well as the relative selectivities of the bonded phases for nitroaniline isomers were determined and discussed. Grushka and Scott [309] used a polyglycine peptide bonded to resin-coated glass beads as well as Porasil and Corasil supports. The attachment of the first glycine was carried out by adding tert-butyloxycarbonyl-glycine and triethylamine to the chlorinated supports. This mixture was refluxed almost overnight in dry EtOH. Elemental analysis showed a butylcarbonylglycine content of 0.0214, 0.1, and 0.293 mmol on Corasil, glass beads, and Porasil, respectively. Ray and Frei [310] developed a brush-type packing material for the separation of polynuclear azaheterocyclics. The material was prepared by bonding p-nitrophenylisocyanate to the surface silanol groups of a porous layer bead. This brush material behaved as a good charge acceptor and this behavior for forming donor-acceptor complexes was affected by the stereochemistry of the molecules to be separated. Of these effects, donor-acceptor complex formation played the major role in the separation process.

A pellicular bonded phase with a polyamide structure, called Pellidon, was prepared by Rabel [311]. This bonded phase is stable to a series of mobile phases from hexane to 0.1 N HCl or 0.1 N NaOH. Because of its polyamide structure, it is able to make H-bonding, thus is a suitable adsorbent for separating compounds with NH_2, NHR, and OH groups. McNair and Chandler [312,313] developed a pellicular packing, Pellisieve, consisting of a thin film of molecular sieve fused onto hard core glass beads. They also modified this packing material by bonding diethanolamine onto its surface. Ross and Jefferson

[314] have described the use of in situ-formed open pore polyurethane as a chromatographic support and evaluated for GC and LC separations. Karger et al. [315] have introduced surface-etched glass beads for HSLC, which are only modified and not chemically bonded packings. The results show that above a 15 ml/min flow rate significant bleeding of the stationary liquid phase occurred even with careful presaturation of the mobile phase. A remarkable review has been recently presented by Locke [316] on chemically bonded phases in HSLC.

Stationary Phase

In separations by LLC, a partition of the sample molecules occurs between the mobile and stationary phases according to Eq. (1). The choice of a stationary phase is dictated by the sample to be separated and is somewhat empirical at the present stage of development. Normal LLC uses a polar stationary phase and a non-polar mobile phase, whereas reverse-phase LLC uses inverse phases. Most LLC separations capable of manipulating the character of both phases can be carried out on relatively few stationary phases. The most widely used solvent pairs are tabulated in Table 2.1.

The optimum level of partitioning-phase loading is largely dependent on factors such as viscosity of the stationary phase, molecular weight, polarity, and peculiarities of the support [98]. The stationary liquid phase must be uniformly coated onto the support, mainly by solvent evaporation techniques [93,317]. Suggested stationary phase loadings are tabulated in Table 2.2. Prepacked columns can be coated by the in situ technique which was introduced in GC by Horvath [318]. By this technique, the liquid stationary phase dissolved in a volatile solvent is pumped through the column, and the solvent is removed by passing dry gas through it. It is recommended to keep the column at elevated temperature by placing it in an oven for some hours to

TABLE 2.1

Most Widely Used Solvent Pairs

Stationary phase	Mobile phase
Normal phase chromatography	
Corasil	Chloroform-hexane Isooctane-iso-PrOH
Durapak OPN	Hexane Hexane-iso-PrOH
Alumina	Acetone Chloroform-iso-PrOH
BOP	Hexane-chloroform CH_2Cl_2 Isooctane-alcohol
Permaphase ETH	Hexane-alcohols Isooctane-alcohols CH_2Cl_2 Chloroform Tetrahydrofuran-alcohols
Reverse-phase chromatography	
Bondapak C-18	Acetonitrile-water MeOH-water Acetonitrile-$(NH_4)_2CO_3$
Permaphase ODS	MeOH-water
Durapak n-octane	Acetonitrile
Poragel	Water-EtOH Acetonitrile-water

ensure the even distribution of the stationary phase [93,319]. Another in situ technique of coating is the steady-state approach in which a mobile phase, presaturated with a stationary phase, is passed through the column [320]. By absorbing a constant amount of stationary phase, an equilibrium state may be achieved on the support. Kirkland and Dilks [321], after pumping a sta-

TABLE 2.2

Suggested Stationary Phase Loadings

Adsorbent	Maximum loading (%)	Minimum loading (%)
Silica gel	-	5
Small-particle silica gel	35	3-5
Zipax	2	0.5
Corasil I	1.5	0.5
Corasil II	2.5	0.5
Pellosil	4.5	1.0
Aluminum oxide	-	3
Silica gel (heavily loaded)	-	45-50

tionary phase dissolved in a solvent through the column, slowly pumped another solvent through it which was immiscible with the stationary phase, but miscible with the first solvent. Under these conditions, the stationary liquid phase is claimed to be homogeneously distributed in the column.

The use of monomers or unique liquid as stationary phase is preferred to polymeric materials or liquid mixtures, although ternary or quaternary systems mixed to form two immiscible phases were successfully used in HSLC [19]. However, as Huber [322] indicated, there are two main practical limitations: (1) the more the stationary phase and mobile phase become similar, the more the stationary phase will be displaced by the moving phase, i.e., the column becomes unstable; (2) the larger the partition coefficient becomes, the less soluble the sample is in the eluent, i.e., the detection of the sample becomes more and more difficult. Horgan and Little [323] compared conventionally coated and chemically bonded phases, using porous and pellicular supports. Column efficiencies were found to be higher on the chemically bonded phases.

Selection of the Mobile Phase

Many different organic and aqueous solvents are available as mobile phase, influencing separation. These should be selective for components. In preparative chromatography, volatility of the mobile phase is desirable. Resolution in a chromatographic separation, as discussed earlier, is dependent on N, α, and k. For a given stationary phase, a strong solvent will cause low k values, whereas a weak solvent will result in high k values, that is, the solute prefers the stationary phase. There are general criteria [324] for the selection of a good mobile phase, such as: the mobile phase should (1) dissolve the sample; (2) keep the column stable [325]; (3) have appropriate viscosity [55, 177]; (4) be pure [326]; (5) be compatible with the detector [192]; and (6) satisfy a number of special criteria (the mobile and stationary phases must be immiscible in one another; active fluorides should be avoided with glass containers; and should not contain dissolved oxygen; etc.).

For HSLC, mobile phases may be listed according to their solvent properties in eluotropic series, which offers a choice in the k values. Using a solvent lower than previously applied in the eluotropic series, the k value of an individual sample component will decrease, and inversely. Usually, there is a maximum solvent strength which can be tolerated for a given stationary phase.

In LSC, water is frequently added to both mobile phase and adsorbent to reduce the strong solute-stationary phase interactions. Snyder [327] has estimated that maximum linear capacity in LSC usually occurs for 50 to 100% of a water monolayer, that is, 0.02 to 0.04 g of water per 100 m^2 of adsorbent. Addition of the right amount of water can raise adsorbent linear capacity to the dry adsorbent by a factor of 5 to 100 [328]. For pellicular adsorbents, 0.5 to 1.0% of water should be added.

The water on the adsorbent must not change during separation. HSIEC is usually carried out in aqueous solution, but sometimes organic solvents are mixed with aqueous buffers to provide unique selectivity changes and improve sample solubility [329-331]. In HSGPC, the solute should interact with the free mobile phase, the adsorbed mobile phase, and the gel matrix to reduce resolution dispersion and skewness of elution curves [332-334]. To improve solute solubility in an HSGPC separation, a mobile phase with lower viscosity should be chosen.

For solving the so-called general elution problem, a change in band migration rates during separation is required [175,176]. The ways whereby this can be executed are gradient elution (or solvent programming), temperature programming, flow programming, and coupled columns [176,179,181]. Among these, only gradient elution is closely related to selection of mobile phases. Comparison of the gradient elution, temperature programming, and flow programming shows that gradient elution is a substantially better general technique whereas temperature programming and flow programming may prove valuable in selected situations [55, 176,179,181,335].

Relations between the Hildebrand solubility parameter and the solvent strength parameter developed for LSC should provide a better guide to choosing mobile phases and predicting solute migration also in HSLC [336,337]. The use of the mobile phase in HSLC is connected with some practical problems, such as solvent degassing and solvent presaturation [338].

2.6 DETECTOR SYSTEMS IN HSLC

Reviews and evaluating studies on the detectors most often used in HSLC frequently appear in the literature [339-351]. HSLC detectors may be classified as being selective or universal, destructive or nondestructive, although it is understood that unfortunately there is not a really universal and nondestructive detector on the market for the time being. Therefore, if samples range widely with respect to their properties to be measured, there is no other choice but to use several detectors at the same time. A selective detector responds to limited classes of compounds. A universal, nonspecific detector should respond to a wide range of compounds and not only to a change in composition of the eluent, but to any fluctuation of the recorded bulk property of the mobile phase. In evaluating the HSLC detectors discussed here, one should keep in mind that their performance depends on the relation between the measured physical quantity and the composition of the effluent as well as on the signal transfer characteristics of the whole detector system. Consequently, if a small amount of sample is measured, the output function is a signal, whose width (which is very important in comparison with the minimum detectable peak width) depends on the sample amount, on the mixing in the sample device (on the physical distribution of the sample in the sampling device), in the connection tube, and in the detector cell, and on the time behavior of the detector.

The primary detector parameters, such as noise drift, absolute and relative sensitivity as well as linearity, and other overall operational parameters are defined and discussed elsewhere [352,353]. Concerning these parameters, it should be emphasized that the usefulness of a detector is strongly limited by its flow and temperature sensitivity, e.g., both microadsorption and refractive index detectors require a high precision control of the temperature and with UV, polarographic, and microadsorption detectors a pulse damping system must be used in order not to transmit a pulsating flow into baseline noise and drift. If detectors are placed in series, they should be connected in the order of increasing dead volume. Using parallel detectors, sensitivity can be lost because of flow and sample splitting.

Light Absorption Detectors

These selective, nondestructive detectors are used mainly in the UV region of the spectrum, although recently there have been several attempts to extend their measuring range over the whole UV and visible spectrum [354-364]. In these detectors, the liquid passes through a flow cell across which light, collimated by lenses from low pressure mercury lamp, is transmitted. The light is filtered or converted into specific wavelengths. High performance UV detectors are commercially available which offer detection at either 254 or 280 nm, and which offer simultaneous detection at both wavelengths [199,365-376]. Monochromators often used in place of the filter provide an infinite selection of wavelengths. UV cells of two different types have been used in commercialized HSLC. In the type of Z configuration, the mobile phase slips half of its pathway along one window of the cell, travels the distance between the two windows against the light beam, and goes out along the other window [343]. To minimize noise and baseline drift owing to the changes in mobile phase flow, an H-shaped cell has also been developed [171]. As the flow inside the cell is not uniform, sometimes turbulence can be generated which tends to scatter the light in the cell, thus affecting the signal. If the absorbance of the mobile phase appears to be changing, a reference cell can be used, although a reference cell can cause more drift and noise than it corrects. Light-absorbance detectors can provide absorbance output which is more convenient than transmittance output because it is directly proportional to sample concentration in accordance with Beer's law. It is desirable for the output of the photocells to be fed to log amplifiers to make the output correctly linear with concentration. Relying on the approximate log function of the photodetector may be useful up to about 0.01 absorbance [349]. Mobile stream monitoring by UV detectors is perhaps the simplest method, for UV absorption characteristics are usually known from previous works and because light-absorbing derivatives can be formed from the compounds with no UV absorption, as in the detection of amino acids [377].

Commercial spectrophotometers could also be easily modified for use as HSLC detectors [190,207,378-386]. In the commercial light absorption detectors for use in HSLC, cell volumes are of 8 to 20 μl, pathlengths of 2 to 10 mm, drift stability ranges 10^{-3} to 4×10^{-4} AU/hr and linear dynamic range is usually of 2×10^{-4} to 1.0 AU. The UV detectors with larger volumes (up to 1.0 ml) are useful mainly for preparative work. Thacker et al. [387,388] developed a minisize UV photometer for continuous monitoring of the effluent at 254 and 280 nm. The photometer operates as a double-beam instrument at both wavelengths indicated and can be used also with gradient elution. A colorimetric detector, with any of the two visible wavelengths, was also developed and integrated with the UV detector into a single modular instrument [388].

UV detectors are frequently used for detection of nucleic acid constituents from ion exchange columns. It is advantageous to carry out these analyses by pellicular materials at temperatures of 70-80°C at which the exchange rate of the resin is sufficiently increased. However, high temperature and high flow rate create excessive detector noise, making the measurement of picomole amounts of these substances impossible. The stability and sensitivity of these detectors can be improved by some modification by which the mobile phase leaves the column, passes through a water-jacketed cooling coil and then enters a water-cooled UV cell [389].

Solvent systems being transparent at the analytical wavelength are preferred for use in HSLC with light absorption detectors. Callmer and Nilsson [390] modified the UV detector used in the Varian 4100 liquid chromatograph. They used Viton foils instead of the original polyethylene foils as gaskets between the silica lenses and the cell. The original Bakelite cell holder was replaced by a brass replica containing cooling coils to achieve better control on cell temperature. These modifications decreased the noise level from 0.002 to 0.0002 AU at a flow rate of 80 ml/min isooctane and allowed detection of less than 0.4 ng of biphenyl (the previous limit was about 10 ng).

Bylina and co-workers [391] applied rapid-scanning spectrophotometry as a new detection system in LC. With this equipment, the spectrum can be periodically scanned, at a rate up to 100 spectra per second. With a mixture of fluorene and carbazole, with an incomplete separation, the changes in the ratio of absorbances at two wavelengths indicated the presence of two components in a single chromatographic peak.

Differential Refractometers

This type is a nonselective, nondestructive detector, whose response depends on the difference between the refractive indices (RI) of the solute and mobile phase. One of the two basic types of RI detectors, the reflection-type detector, is governed by Fresnel's law whereas the other, the deflection type, is ruled by Snell's law. The latter senses the deflection of the light beam, while the first measures the intensity of the reflected light [369,392-399]. Their sensitivity in milligrams of sample is roughly equal to the reciprocal of the difference in RI between solvent and sample [341,344]. The measured physical property RI is, however, affected by the temperature, consequently the sensitivity is also influenced by the stability of the eluent property considered. The RI of water and organic solvents have temperature coefficients of 10^{-3} to $10^{-4}/^{\circ}C$. Therefore, it is of primary importance that reference cells should be used and the temperatures of the two streams should be equalized before entering the refractometer. It is also desirable to thermostat the incoming feedlines to the cell [343]. Johnson et al. [400] manufactured a low-holdup, reflection-type refractometer capable of detecting a change of 10^{-6} RI units with a cell volume of 35 μl.

Deininger, Walkling, and Halász have modified a commercially available RI detector by designing a heat-transfer system with a minimum dead volume in the analytical stream to minimize band spreading [401,402]. Bakken and Stenberg adapted and tested an interferometer as the basis of an LC detector [403]. All commercial RI detectors (except one manufactured by the E-C Apparatus Corp.) have cells with volumes of 3 to 12 μl, permit the use of solvents with RI ranges 1.31 to 1.6 (only the refractometer marketed by Waters Associates, Inc. has a range of 1.0 to 1.75 RI units) and the minimum detectable RI (equivalent to noise) is usually 10^{-7} RI units. Recently, Knauer [404] presented a UV-RI dual detector capable of measuring absorbance and RI in one cell. Using a differential refractometer as HSLC detector, also down-scale peaks may be obtained if the mobile phase is lower in RI than the sample component. More recently, Gow Mac Instrument Co. [405] has manufactured the so-called Christiansen-effect detector whose cell is packed with a solid having the same RI as the mobile phase.

Transport Detectors

These detectors based on the ionization detectors developed for GC operate so that the volatile mobile phase should be removed before the detection process. A moving wire, belt, or chain transports a small amount of the effluent successively into an evaporator, a pyrolizing oven, and the detector. The detector may be a flame ionization (FID), an argon ionization, or an electron-capture detector, although today only the first is in use [406-418]. In the detector reported by Scott and Lawrence [419-421], the sample components are burned to CO_2 and then converted to CH_4 before detection. This modification has resulted in a more sensitive system, especially to oxygenated compounds, with predictable and linear response [422-424]. The sensitivity of this type of moving wire FID has recently been enhanced by a factor of 30 by spraying the column effluent on the wire instead of coating from a block [425]. As a result of this modification the split ratio and consequently the peak areas no longer depend on the mobile phase flow rate. Lapidus and Karmen [426] developed and tested an FID system similar to that of Stouffer et al. [427,428] in which a buffer storage device with a 35-cm-long helix made of nichrome wire moves through the flame. Noise levels were generally less than 5×10^{-13} amp and unaffected by flow-rate changes up to 40 ml/min. The helix speed was suitable between 4.0 and 7.5 cm/min. Pretorius and Van Rensburg [429] developed wires coated with sodium silicate, kaolin, or porous copper in order to improve the performance of the Scott-Lawrence-type moving wire FID. These authors managed to decrease the noise level and enhance sensitivity by a factor of 300 in the lower concentration regions.

Stolyhwo et al. [430] reported an improved FID and associated transport system. In this system, the effluent was added to an endless conveyor consisting of a steel spring with a twisted steel core. After the evaporation of the solvent, the solutes were transported into a pyrolysis burner and pyrolized in a mixture of N_2 and H_2. Sensitivity to triolein was approximately 100 ng. Johnson et al. [431,432] reported a completely destructive FID using an endless belt which transports the entire column effluent into the pyrolizer to increase the sensitivity of the detector. A disk FID reported by Dubsky [433,434] easily transfers the sample with a disk from the column into the combustion space of the detector system. In the detector system, a thin net screen is placed between the electrodes of the FID and the rotating disk, functioning as a Faraday cage and stabilizer of the flow of gases and vapors. The detector signal is reported to be independent of the intensity and changes in the flow rate of the mobile phase.

Heat of Adsorption Detectors

When a sample component displaces solvent molecules, i.e., adsorbs on the packing material surface (or is desorbed) causing concentration change in the composition of the effluent, the heat of adsorption is distributed between the solvent and the adsorbent and raises the temperature by an amount proportional to the change in concentration. Heat loss by the eluent from the column can distort this temperature response [435-437]. Also computer simulation studies suggest that asymmetric response may arise from heat loss from the detector with relatively large volume [438]. Several approaches have been used in detecting the temperature change [439-442]. The detector system developed by Hupe and Bayer is favored and is simple, sensitive, and nonselective [443-445]. In this equipment, the detection and reference cells sensing the temperature change by thermistors are located adjacent to one another in the mobile phase stream, which significantly reduces noise and baseline drift and improves sensitivity by a factor of 10^4 over heat of adsorption detectors having been reported earlier. These detectors have an advantage in that they are ideally suited for determining optimum separation conditions and qualitatively visualizing the elution pattern. Their disadvantage is that the detectors give for each eluted component a positive and negative going peak and the interpretation of two not completely resolved peaks is rather difficult. As this type of detector is subject to any thermal effect, e.g., thermal conductivity and heat capacity of the solvent, it should be maintained at an absolutely constant temperature and at constant flow rate.

These microadsorption detectors can be used with gradient elution only with baseline drift and at reduced sensitivity. The detector is also sensitive to changes in mobile phase flow rate. The dominant flow sensitivity behavior can be attributed to the heating of the solvent by the upstream thermistor and reduced by decreasing the thermistor power dissipation. Munk and Raval [446,447] discussed how this power dissipation could be eliminated. They have improved the operation of the detector either by reducing the bridge voltage or by using higher resistance thermistors or by increasing the distance between the thermistors. They have found another similar approach to achieve a lower flow sensitivity by inserting a metal heat conducting disk between the two thermistor disks. The so-called dual microadsorption detector with two complete detectors in series has also been developed. Smuts and co-workers [448] have also modified a microadsorption detector on an adiabatic approach by continuously adding to the detector an adsorbent slurry so that troublesome negative peaks have been eliminated. The sensitivity of the modified device has reportedly decreased and the modification has increased noise.

Recently, a new type of thermometric detectors has been reported by Gilbert and Dobbs [449]. This detector is based on the volume change caused by the ion-exchange reaction on a strip of ion-exchange membrane placed in the detector. A thermomechanical analyzer can be used to measure the linear dimension change in membrane length as a function of time or effluent volume when the original counterions in the membrane are exchanged for ions in the effluent.

Fluorescence Detectors

This type of detector can measure light reradiated by many compounds including certain metabolites, amino acids, vitamins, steroids, and many drugs excited by UV radiation [450,451]. Fluorescent derivatives of several compounds can also be made [452,457]. A reference cell should be employed to balance fluorescence radiated from the mobile phase, although it must be emphasized that the solvents to be used should be transparent to both the exciting UV energy and the fluorescence wavelengths. The fluorimeter may be very useful as a specific, nondestructive detector with sensitivity as high as 10^{-9} g/ml for strongly fluorescing compounds and is also usable with gradient elution. However, fluorescence suffers from vulnerabilities such as turbidity and quenching. Cassidy and Frei [458] have equipped a commercial fluorimeter with a small volume flow cell. The detector exhibited a very stable baseline and had a noise level equivalent to 2 ppb of quinine sulfate in the 7.5-μl cell. The baseline has shown no dependence on the mobile phase flow rate. An increase in the cell volume to 20 μl has not produced noticeable increase in bandwidth for a 100-cm column, but the shape of the cell was found to be a significant factor with respect to efficency. The detector had a rather slow response time as a disadvantage. Thacker [459] developed a miniature flow-fluorimeter to detect fluorescent compounds in physiological fluids. Scanning fluorescence spectrometry combined with UV detection was employed by Pellizzari and Sparacino [460] for HSLC separations. The column effluent was first monitored by a UV detector, then by a spectrofluorimeter equipped with a microflow cell. A flow-fluorimetric detector was applied by Hatano et al. [461] to monitor HSLC effluents. The spectrofluorimeter is equipped with double-beam optics and a flow cell, 3 μl in volume. The emission and excitation spectra of the components can be recorded by stopping the elution at peak maxima. A detection limit of 10^{-13} g/ml could be achieved for benz[e]pyrene.

Polarographic Detectors

Polarography can be used as the basis of a selective and sensitive detector in HSLC, measuring continuously and recording the current between a polarizable electrode and a nonpolarizable electrode in the function of time. The solution must have electric conductivity and sometimes it is necessary to introduce a suitable electrolyte into the mobile phase before its entering the column. The sample is, therefore, dissolved generally in an aqueous solution of an indifferent electrolyte. When making measurements in LC by polarographic detectors, the design of flow cells is very important. Dropping mercury electrode flow cells were often reported [462-464]. Koen et al. [465] employed a micropolarographic detector with a dropping mercury electrode for the analysis of pesticides. They applied a constant voltage, separated oxygen and other electrochemically active impurities by means of a column, and carried out the measurements in a liquid flow, which resulted in a detection limit of about 2 orders of magnitude (10^{-7} to 10^{-8} M) over classical polarography.

The polarographic detector reported by Joynes and Maggs [466] consists of a carbon-impregnated silicone rubber membrane used as an electrode suitable for oxidation and reduction processes for both organic and inorganic compounds [467]. Reportedly, it gave a sensitivity of about 4×10^{-1} Al/mol for organic and 1×10^{-2} Al/mol for inorganic compounds and had a detection limit of about 2×10^{-9} mol/liter for organics. More recently, MacDonald and Duke [468] have evaluated the use of potentiostatic pulse polarographic techniques in HSLC. Solid-electrode flow cells were used as detectors. They found several advantages over constant-applied-potential techniques, e.g., increased sensitivity (15- to 20-fold), less effect of flow-rate variation, and better electrode stability. This type of detector is difficult enough to use in quantitative work, since the signal may vary with mobile phase, flow rate, voltage, and cell construction.

Radioactivity Detectors

Radioactivity measurements may be very important in the investigations of biological systems. Such detectors in HSLC are not widely employed today, although the technique has a number of advantages such as wide linear range, adaptability to gradient elution, high sensitivity, and suitable quantitation. In this method of detection, the effluent passes through a flow cell or tube packed with a suitable scintillator (the calcium fluoride scintillator is usually recommended) so that the isotope will be in direct contact with the scintillator. The light pulses generated can be detected by a photomultiplier. In a recent work, Schram [469] studied various types of scintillation cells and their efficiencies for ^{14}C and ^{3}H. The packed cells mentioned above have efficiencies of 55% for ^{14}C and 2% for ^{3}H. The mixing system presented by Hunt [470], in which a stream of liquid scintillator miscible with the mobile phase is mixed into the effluent, has efficiencies of 70% for ^{14}C and 30% for ^{3}H. A similar method has been reported by McGuiness and Cullen [471].

A flow cell filled with crystalline 2,2′-p-phenylene-bis-(5-phenyloxazole) as scintillating material, suitable for the continuous measurement of weak β radiation in conjunction with HSLC has been presented by Sieswerda and Polak [472]. The efficiencies for ^{14}C and ^{3}H in the case of an effective fluid volume of 0.1 ml were 51 and 5%, respectively. Heterogeneous and homogeneous scintillation countings for detecting β radioactivity in LC effluents have been compared by Schutte [473]. With heterogeneous scintillation counting, the solution is led through a U-shaped flow cell filled with cerium-activated lithium glass beads, whereas in the homogeneous case, part of the effluent is mixed with a scintillator solution and passed through an empty helical cell. The counting efficiency is reported to be better in the homogeneous counting system. In this case the activity per peak may be as low as 2 nCi for ^{14}C and 5 nCi for ^{3}H, without splitting. The monitoring of both gas and liquid chromatographic radioactive samples in effluents can be conveniently carried out in one apparatus [474]. For high-energetic β or γ radiation, conventional radioactivity measurement methods can also be used in HSLC [475].

Other Detectors

Several other detectors have been employed in HSLC for monitoring column effluents. Their common characteristic is: they were all used in special cases for a certain range of compounds. Of the current monitoring techniques, the electrolytic conductivity detector is one of the simplest to operate and easy to maintain. Here the sample entrained by the mobile phase stream is indicated by a change in conductivity [476,477]. Its sensitivity depends on the difference in conductivity between solvent and sample component. Therefore, this type of detector is useful mainly with aqueous eluent and ionic solutes, e.g., amino acids, without solvent gradient. Commercially available detectors of such type consist of a flow cell with a two- or three-electrode design, through which the total stream of effluent passes. Under appropriate conditions in temperature and flow rate, compounds in a concentration as low as 10 μg are detectable and the detector response may be linear with concentration over a wide range. Pecsok and Saunders [478] reported a conductivity detector. In this detector, the contribution of the strongly reduced dead volume (less than 5 μl) to retention could be neglected. The detector response, however, was nonlinear over the entire range studied because no compensation for different cell capacitance was employed. Tesarik and Kalab [479] described a conductivity detector with a working volume of less than 0.5 μl for use with water as the mobile phase. Minimum sensitivity was 1.14×10^{-8} g/ml of KCl.

Johnson and Larochelle [480] reported the use of a coulometric detector which functioned with virtually 100% efficiency for a large range of flow rates. A detection limit of 1.59 ng of iodide was found. Takata and Muto [481] applied constant-potential coulometry for HSIEC. The newly developed cell had a response time of within 1 sec. The relationship between flow rate and electrolytic efficiency was also investigated. It appeared that this cell could be used for the flow rate of up to 6 ml/min with efficiency over 99.5%. When the polarity of sample component and solvent are widely different, the use of dielectric constant or capacitance detector may be profitable. This type of detector is nondestructive to the sample material, but not recommended with gradient elution, particularly if the gradient causes a change in solvent polarity. Recently, the theory of various possibilities of employing a capacitance detector in LC has been evaluated by Haderka [482-484]. The capacitance detector with a capacitor consisting of two brass cylindrical electrodes reported by Vespalec and Hana [485] was tested for several solutions. The sensitivity in optimal cases was reported to be lower than 10^{-6} g/ml.

Other types of detectors proposed or applied are gas density balance [486], vapor pressure [487], polarimetry [488], light scattering [489], recording balance [490], electrochemical detector [491], and atomic absorption [492]. We should also mention a very recent fact. Mass spectrometry (MS) has come to the forefront also in LC as a method for detection and identification. Since MS is capable of detecting as low as 10^{-11} g of sample, some work has been already done on coupling MS and HSLC [493-498]. In these investigations, only Lovins et al. [494,495] used a nearly on-line coupling via a specially designed interface.

2.7 PREPARATIVE TECHNIQUE IN HSLC

As the difference between preparative and analytical LC lies in whether or not fractions with a sample amount suited for subsequent investigation are gained, we may state that HSLC may be considered preparative in nature. This means that only few different peculiarities are characteristic of preparative HSLC in relation to analytical HSLC. There are two general approaches in HSLC to carry out good preparative separation: (1) scaling up a separation system previously carried out analytically; and (2) combining extraction, LC, and HSLC on a preparative scale. The limiting sample load depends on the type of HSLC to be chosen [499-505]. In HSGPC, the maximum amount of sample that can be chromatographed is the largest injection volume consistent with the required resolution; the maximum concentration is limited by the viscosity of the sample and mobile phase. In HSLLC, HSLSC, and HSIEC, the loading capacity of the column is determined by several factors as discussed earlier. In these separation methods, one may choose two ways for sample concentration: overloading in the first section of the column because of high concentration; or some difficulty in detection due to detector overload or dilute solutions. The results obtained by De Stefano and Beachell [63,64,94], as discussed in Section 1.4, support the choice of the second method. However, below the linear capacity of the column or when separating a sample mixture containing several compounds that easily separate into individual peaks, the first method is also permissible. It should also be mentioned that the weight and volume of the sample to be injected onto the column is also determined by its solubility. It is feasible that the sample should occupy a large initial column volume, i.e., the entire cross section of the top of the column, to utilize the highest available column capacity. In this way, however, the infinite-diameter effect may be eliminated [64]. The results presented by Wolf [506], using a column of 23.6 mm in diameter, confirm this and exclude the infinite-diameter effect as an admissible explanation for the good efficiencies obtained with large diameter columns. In attaching a fraction collector to a preparative column in HSLC, one should keep in mind that the sample may be diluted by a factor of more than 100 as it passes through the column. The design of an automatic fraction collector which meets such requirements is reported by Huber et al. [507]. The liquid from one column can also be divided into fractions by means of a multiport switching valve and collected by a collector or chromatographed on another column [508]. Baker et al. [509] discussed methodology involved in scaling-up from analytical to preparative HSLC and general requirements of sample size for subsequent investigation. For positive identification and subsequent use in research and synthesis, a sample weight over 100 mg has been recommended. An example of the use of preparative HSLC to separate the components of a pyrethrum insecticide extract is also described. System capacity has been increased through use of porous Zorbax small particle silica gel packing material and this increase has been illustrated by the separation of 5 mg of progesterone from impurities, using 0.1% MeOH in CH_2Cl_2 as the mobile phase. Under the same conditions, as little as 10 mg of progesterone can also be collected. On two 50 cm x 23 mm columns packed with Spherosil XOA-400 and connected in series, purification of 1 g of cholesteryl phenylacetate can be carried out using a gradient elution from dry hexane to 0.1% MeOH in CH_2Cl_2. In order to achieve the best purity, a center cut from peaks is recommended to collect.

Chang [510] isolated and identified volatile flavor compounds in boiled beef. By using HSLC, it was possible to collect 8 broad fractions some of which retained the boiled beef flavor. Burtis et al. [511] chromatographed 20 ml of human urine on two Sephadex G-10 columns with different diameter, connected in series, and on a 200 x 1.25 cm anion-exchange preparative column. The four fractions obtained from the gel columns were chromatographed further on the anion-exchange preparative column and another analytical column. At some stages in the synthesis of vitamin B_{12}, individual compounds were purified on preparative scale by HSLC [512-515]. For example, a 200 cm x 2 mm column filled with Corasil II and CH_2Cl_2/MeCN/MeOH (55:24:1) as the mobile phase were used to separate the individual monopropionamides of dicyanoheptamethylcobyrinate. For the separation of 1 g of the mixture, the operating conditions were scaled up to allow runs on a 7.5 m x 24 mm column.

A special technique, recycle chromatography, is often employed also in preparative HSLC [516-520]. This technique, in which the unresolved or partially resolved sample eluting from the column (after flowing through the detector) is backed onto the column to achieve the required resolution, has been put into practice mainly in HSGPC [521-527]. Recycle HSLC contributed also to the elucidation of the absolute configurations of (+)-abscisic acid and violaxanthin [528,529]. The optically active diastereoisomers were separated on a 2.8 m x 10 mm column filled with Porasil T, using a mobile phase of 1% iso-PrOH in hexane at a flow rate of 3 ml/min. Approximately 100 mg of sample was injected. Recently, Nakamura et al. [530] have employed recycle HSGPC to the separation and identification of mixtures of polystyrene and a commercial epichlorohydrin-bisphenol A epoxy resin, respectively. On a column (1.2 m x 20 mm) packed with a highly crosslinked polystyrene gel, individual species of polystyrene from trimers up to dodecamers were completely separated after the seventh cycle and collected for subsequent MS investigations.

We should also mention that the use of preparative HSLC makes possible a combination of HSLC with other chromatographic techniques, such as TLC or GC, and electrophoresis [531].

2.8 AUTOMATION, COMPUTATION

General purpose commercial high-speed liquid chromatographs currently available are not automated. There is a need for automation mainly in routine analyses, in clinical laboratories, and industry, and in laboratories where a great number of samples are analyzed. In the past, routine automatic LC was developed almost solely for amino acid analysis. Such equipments commercialized by several companies have now been reconstructed into HSLC [532-540]. Automation and computation of an HSLC system is possible on different levels: sample injection, column operation and regeneration, identification and collection of sample components, data acquisition, and interpretation. Relatively few articles have been published dealing with automatic sample loading [226,228,541,542]. The samplers designed seem to be useful in low pressure operations. Automation during column operation means mainly an automatic gradient elution system, including also regeneration [186-190,543,544]. Further possibilities are automatic temperature and flow programming and automated column switching during column operation.

Chemical reactions involved in identifying compounds separated on an HSLC column may also be automated. Uziel and Koh [545] constructed an equipment for automatic and selective removal of terminal nucleotide residues from RNAs. A paper tape reading control system was introduced by Marsden et al. [546] for programming sample collection in LC. Automatic data acquisition and interpretation is at the most advanced stage among automation and computational levels of HSLC [547-555]. The state of chromatography computation has been reviewed by Gill [548] and by Derge [549]. Time-share systems have been discussed by Gill and co-workers [550-553]. A specially designed computing integrator for HSLC and GC has been reported by Hettinger et al. [554] and by Gill [555]. Other approaches digitize data on punch tape or card and use large computers for identification and calculation [556-558]. An HSLC-integrator-time share system has been employed to do calculations for the analysis of xylene and polynuclear aromatics separated on a 105 cm x 2.4 mm Carbowax 600 brush-type column, using isooctane as the mobile phase [559]. Chilcote and co-workers [115-118, 560-562] in the Oak Ridge National Laboratory (USA) developed an on-line computer system for identification of chromatographic peaks from body-fluid mixtures separated by HSIEC. Cerimele et al. [563] analyzed urine samples and used a computerized data system for identification and calculation. Vestergaard and co-workers [564-566] developed automated methods for clinical analyses; first introduced a one-channel automatic readout system with punched paper tape output for calculation on a computer; this was later replaced with a high-capacity multichannel system involving a direct on-line computer connection. This pressure-forced system has been applied for the analysis of amino acids and steroids in biological fluids. Recently, in their excellent paper, Barth et al. [567] have reported a study of precision in HSLC, using a dedicated computer.

Chapter 3

APPLICATIONS

A variety of analytical and preparative problems have been solved by means of HSLC and recent developments in this field have given a flexible tool to investigators also for routine analyses [125,568-573]. The examples given here in this chapter are representatives of this flexibility and cover all the branches of the HSLC-separated compounds of interest for chemists and biochemists. Consistent effort has been made to present a complete display as far as possible and we must apologize for the papers which happened to escape our attention.

3.1 ALCOHOLS, ALDEHYDES, AND ACIDS

Aliphatic alcohols are separated usually by GC. As aliphatic alcohols are usually undetectable with the present HSLC detectors, their determination at trace level means a challenge for the investigator. Traces of low molecular weight mono- and diethylene glycols as 3,5-dinitrobenzoates in polyethylene glycols were determined down to 72 ppm [574]. A pellicular packing column, heptane/ethyl acetate (3:1) mobile phase, and UV detection were used. A mixture of aliphatic alcohols containing MeOH,

EtOH, iso-PrOH, n-BuOH, tert-BuOH, and n-amylacohol were separated on a 1-m-long Pellisieve 8AH column, using RI detection and benzene/cyclohexane/ethyl acetate (10:9:1) as the mobile phase [312]. This great selectivity appears to be a combination of pore size effects coupled with hydration of the cations. Aromatic alcohols may be easily analyzed by HSLC. For this purpose, a Zipax pellicular packing column coated with trimethyleneglycol (TMG) was used [575]. Benzyl alcohol, its methyl- and dimethyl-substituted derivatives, 2-phenylethyl alcohol, and cinnamyl alcohol were separated, using hexane or heptane mobile phases saturated with TMG. On a BOP column, benzyl and cinnamyl alcohols had the same retention time, i.e., were not resolved. The compounds except cinnamyl alcohol can be separated on a column packed with 1.5% BOP on Vydac adsorbent [576]. The pellicular silica packing material Pellosil coated with 2% BOP has also been used successfully for separating α-methyl benzyl alcohol, phenylethyl alcohol, and benzyl alcohol [577]. The best resolution for the mixture of the five compounds mentioned above has been achieved on Permaphase ETH chemically bonded packing material [266]. Phenethyl alcohol and benzyl alcohol were separated on a 1.5% BOP-Zipax column, using hexane mobile phase saturated with BOP [174]. On a 2 m x 2 mm column packed with Vydac adsorbent, 2-phenyl-2-propanol, α-methylbenzyl alcohol, benzyl alcohol, and cinnamyl alcohol can be separated, using 1% amyl alcohol in isooctane as the mobile phase and a pressure of 2800 psi [578]. As a test mixture for column efficiency studies, benzene, benzyl alcohol, and benzanilide were chromatographed on a Zipax column coated with 1% triethylene glycol, using hexane as the mobile phase [94]. A mixture of toluene, acetophenone, dimethylphthalate, benzyl alcohol, and phenol was separated on a heavily loaded (50% BOP) column, using n-heptane as the mobile phase [98]. More recently, Kirkland [60] has separated aromatic alcohols by HSLSC on porous silica microspheres (8-9 μm) and by HSLLC on this adsorbent (5-6 μm) coated in situ with BOP, using hexane as the mobile phase. On a similar column in HSLSC, nitrobenzene, methyl benzoate, α,α'-dimethyl benzyl alcohol, and cinnamyl alcohol were separated, using a linear gradient at 10%/min from 0.1% iso-PrOH in hexane to 1.0% iso-PrOH in CH_2Cl_2 [579].

The 2,4-dinitrophenylhydrazones of the C_1-C_7 aliphatic aldehydes were separated on a pellicular packing column, using heptane/ethyl acetate (97:3) as the mobile phase [574]. HSLC using Permaphase ETH chemically bonded packing was proposed by Papa and Turner [580] as a method which could be used to separate and determine carbonyl compounds as their 2,4-dinitrophenylhydrazone derivatives. The sensitivity limit was about 5 ng. Application of the method for the determination of carbonyl compounds in automobile exhaust was cited. Benzaldehyde and p-nitrobenzaldehyde were separated from aromatic amines and nitro derivatives by HSLSC using isooctane/$CHCl_3$ (9:1) as the mobile phase [576].

Ascorbic, dehydroascorbic, and diketogluconic acids were separated by a pressure-forced anion-exchange system [581]. The column, 56 x 0.9 cm, was developed by 0.05 M H_3PO_4 at a flow rate of 35 ml/hr. The isomeric fumaric and maleic acids can be separated within 2 min on a Zipax SAX column, using 0.01 N HNO_3 as the carrier [230]. The separation of mono-, di-, hydroxy-, and ketoaliphatic carboxylic acids by HSIEC was reported by Kaiser [582] (Fig. 3.1). A column of 50 cm x 2 mm packed with Aminex A-14 resin and a mobile phase of 10 mM $NaHSO_4$ and 5 mM HNO_3 in H_2O were used. The influences of pH, ion concentration, and ion type in the mobile phase are discussed.

Nitro-substituted aromatic carboxylic acids were successfully chromatographed on a 1 m x 2.1 mm column filled with water-saturated Pellisieve 8A, using tetrahydrofuran as the

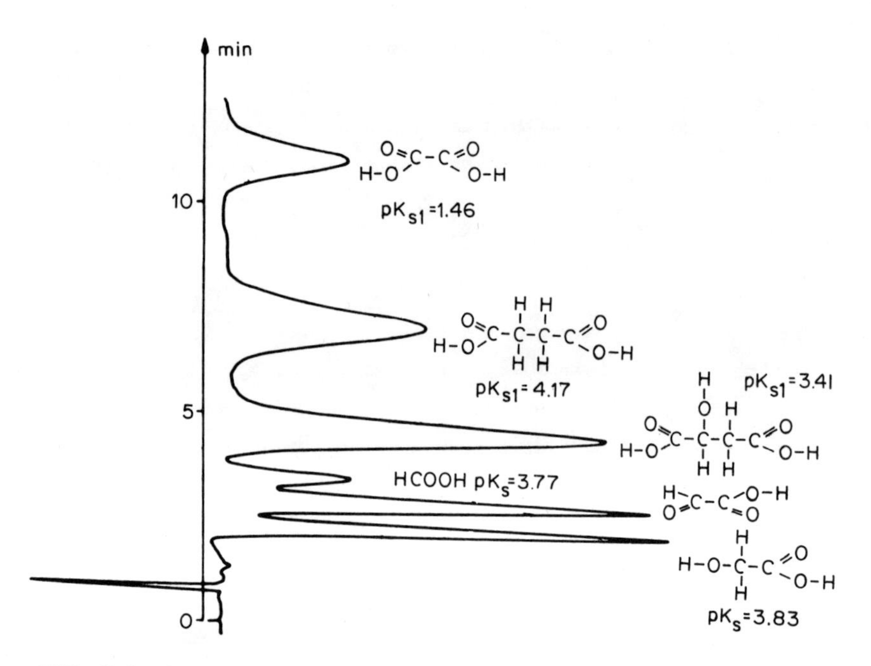

FIG. 3.1 Separation of mono-, di-, hydroxy-, and ketocarbonic acids. Column: 50 cm x 2 mm, Aminex A-14, 17-23 μm; mobile phase: 10 mM $NaHSO_4$ and 5 mM HNO_3 in H_2O; flow rate: 45 ml/hr. (From Ref. 582.)

mobile phase [312]. Homologous N,N-dimethyl-p-aminobenzene-benzoyl esters were separated in a pressure-forced system on a 24 cm x 2.7 mm Dowex 50W-X2 column, using an eluent of 0.925 M HCl in 80.5% EtOH at a flow rate of 1 ml/min [583]. Quantitative analysis of several benzoic and cinnamylic acids was carried out by pressure-forced LC at low pressure on a Polycar column [584]. Hövermann et al. [585] carried out the separation of phenolic acids on a small-particle (5-8 μm) silica gel column, using CH_2Cl_2/EtOH/H_2O (85.2:12.5:2.3) as the mobile phase. The cis-trans isomers of p-cumaric acid as well as ferulic acid were separated. Separation and determination of 10 benzene-polycarboxylic acids were carried out on a strong anion-exchange column, using a linear gradient of 0.01 M to 0.1 M $NaNO_3$-H_3BO_3

at 5%/min (pH 9.1) [586]. The quantitative analysis of these acids as well as o-aminobenzoic acid and methylmaleic acid was achieved by using internal standards. Dehydroacetic acid and 4-hydroxy-2,6-dimethylnicotinic acid were separated from substituted pyridinol and pyridone compounds on a Dowex AG1-X2 column (50 cm x 2.8 mm) [587]. An acetic acid/MeOH gradient and a specially designed gradient elution apparatus were employed.

Benzoic, toluic, and terephthalic acids were separated on a column filled with Zipax SAX resin, using a pH 9.2 borate buffer and a column pressure of 1200 psi [588]. On the same column, dye intermediate naphthalene sulfonic acids (Schaffer's, R, and G salts) could also be separated using 0.025 M sodium nitrate as the mobile phase. Nitriloacetic acid was determined in an aqueous solution of detergents by HSIEC [589]. The mobile phase was 0.02 M $Na_2B_4O_7$ at a pH of 9.0. 2,4-Dichlorophenoxyacetic acid can be separated from its iso-Pr-, isoBu-, and ethylhexyl esters on a reverse-phase column, using MeOH/H_2O (3:2) mobile phase at a temperature of 60°C and a flow rate of 2.2 ml/min [265].

The separation of tricarboxylic acid cycle intermediates and related carboxylic acids was studied by Stahl et al. [590], using silica gel (15-25 and 25-28 μm) columns and a two-phase system of 0.1 N H_2SO_4/$CHCl_3$/tert-amyl alcohol. The sample was dissolved in the liquid stationary phase. The separation of fumaric and pyruvic as well as lactic and succinic acids was not complete. A similar incomplete separation was observed between citric and isocitric acids. α-Ketoglutaric acid always showed three peaks probably due to instability. Prior et al. [591] applied a pressure-forced gradient system (from 200 to 300 psi) and a silicic acid column for quantitating effects of fertilization on plant organic acids. Seventeen acids were separated by using a complex gradient elution system of tert-amyl alcohol/$CHCl_3$.

Katz and Pitt [592] employed a cerate oxidation fluorimetric method in monitoring organic acids (lactic, pyruvic, malonic, α-ketoglutaric, 2-furoylglycine, 5-hydroxymethyl-2-furoic) in urine. This method relies upon the reduction of reagent cerium(IV) to the fluorescent cerium(III) by compounds in the eluate. Benzene and naphthalene sulfonic acids were chromatographed on a Zipax SAX column at 60°C [260]. The initial eluent, 0.0025 M $HClO_4$, was changed to 0.005 M $HClO_4$ after the separation of the benzene sulfonic acids to elute the last component, sodium naphthalene-2-sulfonate, more rapidly.

3.2 ALKALOIDS

As a number of alkaloids are of biological importance, an improved method over the classical methods of analyzing these compounds is valuable. Snyder [196] developed an HSLSC method for rapid qualitative and quantitative analysis of hydrogenated quinoline mixtures. The separation was carried out on an alumina column, using water-saturated solvent prepared by passing 35% CH_2Cl_2/pentane through a column of 30% water/silica. A CH_2Cl_2 gradient elution was employed, varying the CH_2Cl_2 concentration from 25 to 60% by volume. Also, adsorption energies of some alkaloids were measured on acidic, neutral, and basic aluminas [593]. Mixtures of 10% diethyl ether and 10% CH_2Cl_2 with n-pentane were used for elution. The results show that the application of n-pentane/10% CH_2Cl_2 as the mobile phase on acidic alumina is advantageous. Ipecac alkaloids, Emetine and Cephaeline, were chromatographed on a cyanoethyl silicone column by using n-heptane mobile phase, whereas *Veratrum viride* alkaloids could be separated on TMG or Carbowax 400 stationary phases [594]. Three major purine alkaloids, caffeine, theobromine, and theophilline commonly occurring together, were separated successfully on Corasil coated with 1.1% poly-G-300 stationary phase, using heptane/EtOH (10:1) as the mobile phase [595]. Also on this column, strychnine and brucine could be determined.

Cation-exchange chromatography was also applied successfully for the separation of pyridine, quinoline, isoquinoline, and 8-hydroxyquinoline with 0.15 N $NaNO_3$ mobile phase [596]. Quinoline and its 8-hydroxy derivative reversed elution at a concentration of about 0.11 M $NaNO_3$. Separation of substituted benzimidazoles on a cation-exchange column was presented by Kirkland [230]. The compounds were baseline-separated in about 15 min.

Talley [597] reported an HSIEC method for the separation of the cyano-, carboxamido-, and carboxy-substituted pyridine isomers achieving quantitative results. Resolution was not sufficient, however, to permit baseline separation of all nine isomers in one chromatogram. The isomeric picolines and pyridine were also separated on a column packed with 1% BOP. HSLC separations of polynuclear azaheterocyclic compounds on a column filled with silver ion impregnated porous layer bead were reported [598]. The compounds were separated by donor-acceptor complexing between the silver ions and heterocyclic nitrogen atoms. The elution order was influenced not only by the basicity of the compounds but steric accessibility of the nitrogen lone-pair. Were basicity the only factor, the elution order would follow the order of increasing pK_a. The results indicate that the elution order departs from this rule and does not follow the trend of increasing pK_a. Ray and Frei [310] achieved a better separation of these compounds.

An accurate determination of morphine as active ingredient in a suppository can be carried out on an alumina column, using $CHCl_3$/cyclohexane (7:3) as the mobile phase [599]. Separations of four opium alkaloids (papaverine, thebaine, codeine, and morphine), cinchona alkaloids (quinine, cinchonine), morphine, heroin, caffeine, and cocaine were reported by Wu et al. [600, 601]. They used columns packed with pellicular adsorbents coated dynamically with the poly-G-300 stationary phase. Quantitative analysis of three rapidly eluting opium alkaloids was also described, at nanogram up to microgram levels. In addition to the separation of the four opium alkaloids mentioned above, cryptopine and narcotine can also be separated on a strong cation-exchange column, using an aqueous 0.77 M NaOH/boric acid (pH 9.8) mobile phase at a linear velocity of 0.75 cm/sec [602]. Under the same conditions, except for 0.15 M NaOH, separation of morphine, 6-(O-acetyl)morphine, and heroin can be achieved (Fig. 3.2). A study of the parameters influencing the chromatographic behavior of these alkaloids and methadone has been recently reported [603].

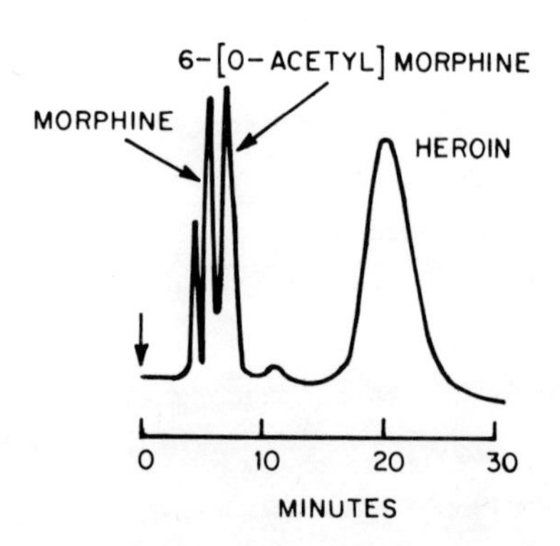

FIG. 3.2 Separation of alkaloids. Column: 1 m x 2 mm, Zipax SCX; mobile phase: 0.04 M NaOH, boric acid added to pH 9.3, 12% CH_3CN, 2% n-propanol; linear velocity: 0.44 cm/sec. (From Ref. 602.)

A preliminary study of the separation of 13 oxindole alkaloids on pellicular silica columns was reported using a mobile phase of MeOH/H_2O (4:1) at 60°C [604]. Ergotamine, ergocristine as well as some other ergot alkaloids were separated on Corasil columns and detected by a fluorescence method [605]. A mixture of $CHCl_3$/MeOH/AcOEt/AcOH (60:20:50:3) was used as the mobile phase. Tropane alkaloids (tropine, pseudotropine, 3,6-dihydroxytropine, and teloidine) have been recently separated by Rajcsanyi [606] on a new small-particle bonded phase. Other tropane alkaloids (scopolamine, atropine, and hyosciamine) were separated on a silica gel column, using NH_4OH/tetrahydrofurane as the mobile phase [607].

3.3 AMINES, AMINO ACIDS, AND AZO COMPOUNDS

Ion-exchange resins have been used for both simple and complex analyses of basic amino acids and related compounds. The procedure has been automated in amino acid analyzers. Cation-exchange resin can be well employed in separation of amino acids by HSIEC [268]. The mobile phase is 0.1 M ammonium acetate to which HCl should be added until pH is 2.5. The reported results show that tyrosine and phenylalanine were not completely separated. Eighteen amino acids were, however, determined by ion-exchange separation, as reported by Inglis and Nicholls [608]. A similar separation was achieved by Ertingshausen et al. [538] using small-particle cation-exchange resin and citrate buffers in an automated system. In a fully automated amino acid analyzer, 19 amino acids can be separated within 210 min [267,609]. A mixture of 18 amino carboxylic acids and other amines and amino acids (taurine, urea, citrulline, glutamine, α-amino-n-butyric acid, ornithine, and tryptophan), as the known composition of plasma, was analyzed by using sodium and lithium citrate buffers in a gradient elution system [610,611]. In the separation, taurine and urea were poorly resolved; threourine, glutamine, and serine, as well as proline and citrulline appeared as single peaks. An improvement in the separation of the threourine/glutamine/serine complex was followed by an inadequate separation of α-amino-n-butyric acid and leucine and an undesirable longer separation time. Liao et al. [612] used a single column (60 cm x 2.8 mm) packed with a small-particle porous ion-exchange resin to analyze a standard solution of 18 amino acids containing 3.5 nmoles of each amino acid.

A recently developed fluorescence amino acid analysis was used by several investigators [455,613-619]. This procedure applies a new reagent, fluorescamine, to determine amino acids, peptides, proteins, and other primary amines at the picomole level (Fig. 3.3). Roth and Hampai [620] utilized a reaction of o-phthalaldehyde and 2-mercaptoethanol with all α-amino acids. This reaction can be used to monitor amino sugars separated by HSLC. Dansyl amino acids can be separated on a Zipax column, using methylethylketone/light petroleum (1:20) as the mobile phase and determined by a fluorescence detector [454]. The determination of a mixture of five ^{14}C-labeled amino acids has been described by Sieswerda and Polak [472] using a radioactivity detector. The peaks of the amino acids in the chromatogram represented a radioactivity of 6 nCi each. Histidine, phenylalanine, tyrosine, and tryptophan could be separated on a polypeptide-bonded phase column, using H_2O as the mobile phase [309]. Tyrosine and phenylalanine can be separated on a Pellionex SCX column at 40°C using 0.0025 M $NH_4H_2PO_4$ (pH 4.5) as the mobile phase [621].

Sixteen phenylthiohydantoine (PTH) amino acids were separated on a two-column system by HSLC [622]. A small-particle silica gel (5 μm) column (50 cm x 3 mm) and CH_2Cl_2/dimethylsulfoxide/tert-BuOH as the mobile phase were used to separate

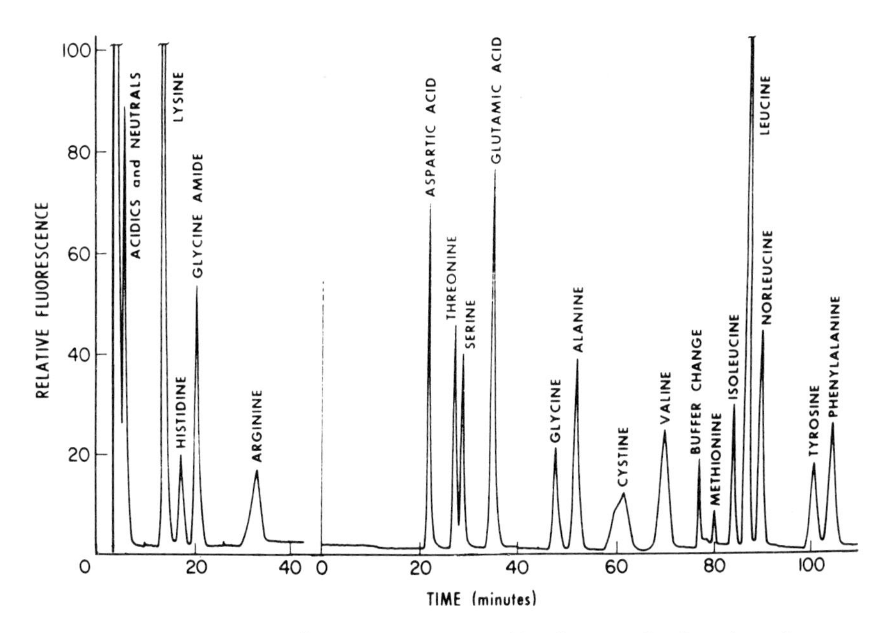

FIG. 3.3 Chromatography of amino acids in a hydrolyzate of 2.5 μg of bovine serum albumin. Column: 30 cm x 2.8 mm, Durrum DC-4A; mobile phase: a complex gradient of sodium citrate; flow rate: 7 ml/hr; detection: fluorimetric with fluorescamine. (From Ref. 616.)

the "hydrophobic" amino acids. For the separation of the "hydrophilic" amino acids, a mobile phase of CH_2Cl_2/DMSO/H_2O (80:15:2) was employed.

Thirteen PTH amino acids in three separate groups were chromatographed by Graffeo et al. [623] using a small-particle silica gel column and a complex gradient of $CHCl_3$/heptane. The separation of glycine and all peptides from diglycine to heptaglycine was achieved on a cation-exchange column, using a 0.2 M sodium citrate buffer, pH 3.35 [624]. The results show that the resin also acts as a porous gel (for peptides with higher molecular weight) and thus separates the peptides not only according to their pK_a values but also on the basis of their molecular size. Thyroidal iodoamino acids can be separated rapidly by chromatography on columns of rigid CPG packing, using a mixture of ethyl acetate/MeOH/2 N aqueous NH_4OH (40:10:4) as eluting solvent, although the separation was classical rather than high-speed chromatography [625]. Phenethylamines were chromatographed on an ion-exchange column, using 0.2 M $NaNO_3$ (pH 3.15) as the mobile phase [626]. During the analysis temperature was 55°C.

A systematic survey on the elution of anilines from two brush-type (Durapak OPN and Durapak Carbowax) and one adsorbent (Corasil II) columns has been presented by Sleight [627] to gain an understanding of the mechanisms causing retention and the structural factors modifying it. Sleight has well utilized the earlier experiments on the separation of anilines [93,102, 274]. The retention caused by any direct interaction of the substituent with stationary phase due to its electronically induced effect on amino groups was examined with some monosubstituted benzenes. The results do not show clear correlation with constants in the Hammet plot, although they suggest that electron-donating substituents cause increased retention. Retention can be decreased by introducing ortho substituents into anilines. N-substitution and di-N-substitution decrease retention owing to a decrease in capacity factors. These results indicate that hydrogen bonding does not play such an important role in aniline retention as it does in phenol retention.

On TMG or BOP columns, p,p′-methylenedianiline can be purified from its impurities, using heptane as the mobile phase [575]. The same purity can also be achieved by using small-particle silica gel and a mobile phase of 0.7% MeOH in 75% H_2O-saturated CH_2Cl_2 [579]. The separation of aniline, N-ethyl-aniline, and N,N-diethylaniline can be carried out on a column packed with small-diameter (10 μm) silica gel coated with 33% BOP [628]. HSLSC has been successfully used for the determination of nitroanilines [576] as well as isomer xylidines [629], using pellicular silica gel packing materials. Nitroanilines can be separated also on alumina packing material, using benzene diethyl ether or 40% CH_2Cl_2 in hexane as the mobile phase [319,630]. A polar-bonded phase [307] and 0.5% iso-PrOH in heptane mobile phase as well as a Durapak Carbowax 400 brush-type bonded phase and $CHCl_3$ [198] may be useful in the separation of the nitroaniline positional isomers. Aromatic amine antioxidants (N-ethylaniline, N,N-diethylaniline, diphenylamine, and N-phenyl-2-naphthylamine) were separated by Majors [631] using pellicular packing materials coated with 0.5% BOP as well as Durapak OPN bonded phase and isooctane as the mobile phase. The elution order was the same on the columns. Selectivity for each solute was the greatest on the brush-type packing material; resolution and the H values were, however, less favorable. Kiselev et al. [87] showed that the mixture of diaminophenyl-methane isomers could be completely separated on an LSC column (aluminum oxide), whereas there was no complete separation by GLC.

O-Toluidine and m-toluidine can be separated on a Pellumina column, using 1-chlorobutane as the mobile phase [632]. The phenetidine isomers can be separated on a 1.0 m x 1.8 mm Pellidon column using 2.5% acetic acid in hexane as the mobile phase at a flow rate of 60.4 ml/hr [311]. A mixture of o-, m-, and p-sulfanilic acids can be separated on a pellicular weak anion exchange resin at 55°C, using 0.002 M KH_2PO_4 (pH 2.8) as the mobile phase [633]. A complex mixture, including azobenzene, 1- and 2-naphthalene amine, 2-methylaniline, quinoline, and isoquinoline was successfully chromatographed on an 8.8 cm x 3 mm silica gel (8-12 μm) column, using isooctane as the mobile phase at a flow rate of 6 ml/min [634].

A comparison between HSLSC and TLC in the separation of a lubricant containing also dioctyldiphenylamine and octyl-2-naphthylamine shows the advantages of HSLSC over TLC [635].

Porapak Q and T as packing materials were used for the separation of freon, iso-PrOH, and N,N-diethyl-m-toluamide which were the basic compounds of sprays [226]. The phenylenediamine isomers can be separated on a BOP column, using 10% EtOH in hexane as the mobile phase [636]. However, a Permaphase ETH-bonded phase column gives a better resolution, using 5% MeOH in cyclopentane as the mobile phase [254]. Aniline, 2-chloroaniline, and 2,5-dichloroaniline were separated on a large diameter column (49.3 cm x 10.9 mm) packed with 1% triethylene glycol on Zipax, using n-hexane as the carrier liquid [94]. 2-Chloroaniline, 3-chloroaniline, and 3,4-dichloroaniline could be separated on small-particle silica gel (5-6 μm) coated in situ with BOP, using the same mobile phase as in the above experiment [321]. Azobenzene and substituted N,N-dimethyl- and N,N-diethylazoanilines were separated using 10% CH_2Cl_2 in hexane as the mobile phase [637]. A solvent program of 5-30% CH_2Cl_2 in hexane at 4%/min may improve the separation.

A mixture of biogenic amines, containing also dopamine, 6-OH-dopamine, 5-OH-tryptamine, 5-OH-tryptophan, and L-dopa has been recently analyzed by Kissinger et al. [491] who used a pellicular SCX resin and an electrochemical detector. The detector is a working electrode set in the effluent emerging from the column, held at a fixed potential to oxidize (or reduce) electroactive compounds passing by. Ion-exchange chromatography and a fluorimetric detection method were used to determine urinary polyamines by Veening et al. [638].

3.4 AROMATIC HYDROCARBONS AND SUBSTITUTED POLYNUCLEAR AROMATICS

Although GC is very useful in the separation of hydrocarbons, some molecules have low volatility owing to their high molecular weight so that HSLC may be a valuable complementary method in this field of analysis, too. A classwise separation of a premium gasoline into saturates, monoaromatics, and diaromatics by HSLLC was reported by Stevenson [639]. The elution times of alkylated benzene compounds vary systematically with the number of paraffinic substituents. Increasing number of exocyclic carbon atoms causes decrease in retention time along a straight line. Extrapolation of this behavior indicates that pentadecylbenzene would emerge with the saturated hydrocarbons on a Carbowax 600 column. Naphthenic substitution of the benzene ring apparently increases retention time. Liquid fuels can be divided into four classes by HSLSC: (1) saturated hydrocarbons; (2) monoaromatics; (3) naphthanic hydrocarbons; and (4) biphenyl hydrocarbons [640]. The use of a Porasil T column and heptane mobile phase at a flow rate of 1 ml/min is recommended. Separation of aromatic hydrocarbons, paraffins, and olefins as well as diolefins was achieved by Locke [22] who used squalane as stationary phase and CH_3CN/H_2O as the mobile phase. The separation of six polycyclic aromatic hydrocarbons was carried out on an 8.1% dimethylformamide/Carbowax 350 column with isooctane as the carrier phase [641]. Phenanthrene is not resolved from 7,12-dimethylbenzo[a]anthracene and benzene as the first peak may not be well detected.

Benzene, naphthalene, anthracene, fluoranthene, 1,2-benzanthracene, and 4,5-benzpyrene can be separated on silica gel coated with 40% Fractonitril, using n-heptane as the mobile phase [96]. The same mixture, except for 4,5-benzpyrene, can also be separated on a silica gel column coated with 40% BOP, using isooctane mobile phase [173].

The separation of eight polyaromatic hydrocarbons was achieved by Karger et al. [642], using a 2,4,7-trinitrofluorene impregnated Corasil column and heptane as the mobile phase. HSLSC is also worthy of consideration for the determination of polynuclear aromatic hydrocarbons. Rapid baseline separations can be achieved both with pellicular silica and with neutral alumina packing materials, partially deactivated with water [198,643-647]. A change in these packing materials causes a reverse order of elution of benzanthracene and o-quaterphenyl. In the case of alumina, by changing the solvent, pure hexane for hexane/isopropyl ether (60:40%), all components are eluted sooner and benzanthracene is eluted last, after m-quaterphenyl. Karr et al. [648] used an alumina column and cyclohexane as the mobile phase in a pressure-forced system to analyze aromatic hydrocarbons from pitch oils. The selective separation of aromatic hydrocarbons and their hydrogenated derivatives (altogether 20 compounds) was carried out on Porapak T [649]. A comparison between TLC and HSLSC indicated that more reproducible separations could be achieved by HSLSC.

The separation of a mixture containing toluene, naphthalene, anthracene, fluoranthrene, benz[a]anthracene, benz[a]pyrene, and benz[e]pyrene was demonstrated by Strubert [650]. The limit of detection was about 2 ng when using a Sperosil XOA 400 (4-8 μm) column, isooctane as the mobile phase, and UV detection. HSLC was shown by Stewart et al. [79] to be superior in performance to TLC in the separation of pyrene, fluoranthene, and benzanthracene.

Seventeen aromatic, substituted aromatic, and polynuclear aromatic hydrocarbons were separated by Snyder and Saunders [177] who used an optimized solvent programming system. In comparing normal (isocratic) elution, coupled-columns, solvent, flow and temperature programming, Snyder [176] separated a mixture of tetralin, naphthalene, acenaphthalene, phenanthrene, fluoranthene, 1-methoxynaphthalene, 2-methoxynaphthalene, o-nitrotoluene, 1-nitronaphthalene, and m-bromomethylbenzoate to evaluate these separation techniques. Klimisch [651,652] employed cellulose acetate columns to separate coronene and benzpyrene isomers; the mobile phase was EtOH/CH_2Cl_2 (2:1). Isomeric benz[a]pyrene and benz[e]pyrene could also be separated on a polyamide column [652]. Preliminary TLC results of such a separation could be easily transferred to HSLSC. The same separation of the benzpyrene isomers was also achieved by Ives and Giuffrida [653]. In addition to the cellulose acetate, they also used Durapak OPN and isooctane mobile phase to separate 18 polycyclic aromatic hydrocarbons. On the Durapak OPN column, chrysene and benzo[b]fluoranthene, and benzo[a]anthracene could be entirely separated. The elution order of the hydrocarbons on cellulose acetate was significantly different from that obtained on Durapak OPN. Loheac et al. [654] gave a comparison of different packing materials in the separation of polynuclear aromatic hydrocarbons. They used silicic acid, Spherosil XOB 075, and neutral alumina columns. The effect of deactivation on retention was also studied.

Popl et al.[655] used water-deactivated alumina for the separation of alkylbenzenes with chainlengths from ethyl to n-decyl. The use of methylcyclohexane as a stronger solvent instead of pentane substantially decreased the differences among retention volumes of individual n-alkylbenzenes. Nonpolar solutes such as hydrocarbons may be separated effectively by reverse-phase chromatography, i.e., using nonpolar liquid stationary phase and polar carrier liquids. The most useful mobile phases for this purpose are water/alcohol mixtures, although water/dioxane and water/butyl Cellosolve may also be employed. The resolution and retention of polynuclear aromatics are ruled by the alcohol percentage. At 90% alcohol content in the mobile phase, all aromatics elute as a single peak which gives a method for the rapid estimation of total aromatics present in mixtures [265,656,657]. Telepchak [658] investigated the possibility of reverse-phase adsorption on natural diamond having a surface similar to the inert, nonpolar bonded phases. Benzene and anthracene were separated on a 25 cm x 3 mm column, using MeOH/H_2O (3:7) as the mobile phase.

The recently developed chemically bonded reverse-phase packing materials have substantially improved the efficiency and resolution for these compounds [266,373,643,659-661] (Fig. 3.4). In addition to these compounds, chloro-, bromo-, iodobenzenes, and mixed halogen derivatives of aromatics have also been chromatographed with these materials. Schmit et al. [662] chromatographed auto exhaust condensate fractions obtained by HSGPC. They used a Permaphase ODS reverse-phase column and a gradient elution of MeOH/H_2O (1:4) to H_2O at 60°C. The effect of temperature on efficiency was also evaluated. Bondapak C-18 reverse-bonded phase packing material was employed in the separation of polynuclear hydrocarbons in used engine oils [663]. This material has advantage over those employed in LSC in producing chromatograms which can be used as fingerprints. Sleight [664] studied the chromatographic behavior of 10 alkylbenzenes and 21 polynuclear aromatic hydrocarbons on Permaphase ODS reverse-phase packing material, using MeOH/H_2O mixed solvents as the mobile phases. An increase in retention with increasing number of aromatic rings in a molecule was noted. A simple linear relation was found between log k and the number of carbon atoms in an aromatic hydrocarbon; another linear relation was suggested between log k and the total number of alkylcarbons in alkyl-substituted aromatics. Benzpyrene isomers were also separated by Ledford et al. [665] who used a combination of Durapak OPN and glass beads treated with octadecyltrichlorosilane and a fluorimetric determination. The identification of seven polyaromatic ring systems in petroleum by GPC has been reported by McKay and Latham [666]. Pressure-forced GPC separated the pericondensed aromatic ring systems from the catacondensed systems. Anthraquinones can also be separated on a Corasil packing coated

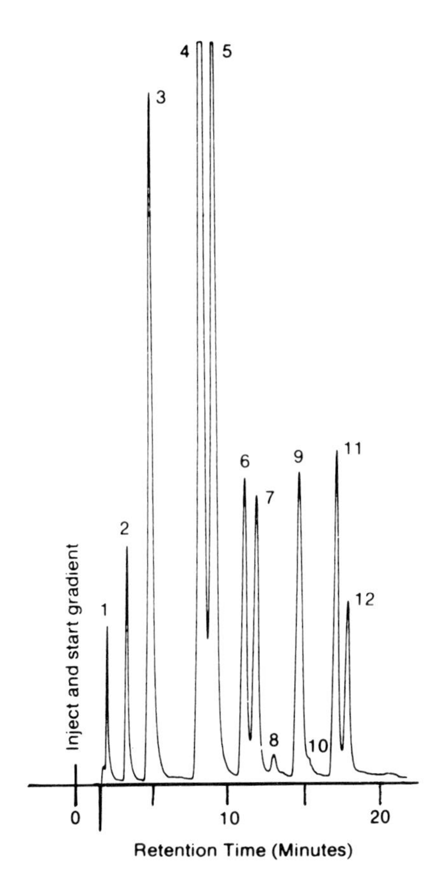

FIG. 3.4 Fused ring aromatics. Column: 1 m x 2 mm, Permaphase ODS; mobile phase: linear gradient from MeOH/H_2O (1:1) to MeOH; temperature: 50°C; flow rate: 1 ml/min; sample components: (1) benzene, (2) naphthalene, (3) biphenyl, (4) phenanthrene, (5) anthracene, (6) fluoranthene, (7) pyrene, (8) unknown, (9) chrysene, (10) unknown, (11) benz[e]pyrene, (12) benz[a]pyrene. (From Ref. 266.)

with 1.3% squalane, using MeOH/H_2O (1:1) as the mobile phase [667]. Studying the effect of temperature on separation showed the performance of the column to be better at 75°C than at 30°C. Generally speaking, increasing the temperature reduces retention volumes and thereby increases the speed of separation. In the above separation, an increase in temperature decreased the partition ratios of the anthraquinones. The advantage of reverse-phase chromatography can be extended to more polar solutes such as anthraquinones by increasing the concentration of H_2O in the mobile phase [668]. Diphenylsulfonates and other aromatic sulfonates can be separated on ion-exchange columns using $H_2O/CH_3CN/$ MeOH (1:1:3) as the mobile phase [669].

3.5 CARBOHYDRATES

Classical TLC and paper chromatographic methods used to determine carbohydrates are time-consuming and only semiquantitative. Classical IEC has been successfully used to separate carbohydrates but analysis, sometimes, requires as long as 12 hr [670]. Thus pressure-forced and HSLC has a great future in the analysis of carbohydrates [671-673]. Kesler [674] used an automated HSIEC system to separate a 15-component solution on an anion-exchange column. A complex gradient was formed from H_3BO_3 and NaOH solutions. An 8-component (sucrose, cellobiose, rhamnose, mannose, arabinose, galactose, xylose, and glucose) standard mixture was used for quantitative calibration. The increase of column temperature enhanced resolution and shifted the elution bands to longer elution times at a given flow rate. The column effluent was monitored for carbohydrates by using the orcinol colorimetric method.

In separations of monosaccharides on ion-exchange resins in their lithium and sulfate forms, aqueous EtOH may also be employed as the mobile phase [675]. Pesek and Frost [676] who used a 95% $EtOH/H_2O$ mixture and pellicular anion and cation exchangers indicated that no retention was observed for unsubstituted sugars (fructose, galactose, sucrose, and 3-O-methyl-D-glucose) showing that the previously proposed partition mechanism for sugars in this mobile phase was correct. Schneider and Lee [677] have recently separated anomeric arylglycosides and 1-thioglycosides by pressure-forced cation-exchange chromatography at 56°C, using water as the mobile phase at a flow rate of 1 ml/min. Phenyl-, p-nitrophenyl-, and o-nitrophenyl derivatives of 20 glycosides were chromatographed and determined by the orcinol method. The anomeric pair of o-nitrophenyl-D-galactopyranosides as well as phenyl-D-galactopyranosides was only incompletely separated. The β-anomers precede the α-anomers; p-NO_2 as well as 1-thio substitution increases retention time and improves separation. Replacement of the -CH_2OH group in sugars increases retention, whereas replacement with -H group decreases the efficiency of anomeric separation despite the increase in retention time.

Nanomole amounts of reducing glucose oligomers (G_1-G_9) were determined by the alkaline ferricyanide and the Park-Johnson methods after chromatography on a Bio-Gel column [678]. The mobile phase was H_2O at a flow rate of 7.5 ml/hr. Hobbs and Lawrence [679] analyzed mono- and disaccharides (rhamnose, fucose, ribose, xylose, arabinose, mannose, glucose, galactose, fructose, sucrose, maltose, and lactose) on a strong cation-exchange column at 65°C, using 85% EtOH/H_2O as the mobile phase. The triethylammonium form of the resin was found to give the best separation. Quantitation was achieved by a moving-wire FID system. The effects of increasing water content in the mobile phase and increasing temperature reduced the capacity ratios, especially for disaccharides. A disadvantage of the method is some ketose decomposition. (See Fig. 3.5.) Oligosaccharides of the dextrose series in a starch hydrolyzate could be separated on an anion-exchange column, using flow programming from 0.12 to 4.2 ml/min of the mobile phase (H_2O/ethyl acetate/iso-PrOH) [298]. The separation was completed within 55 min.

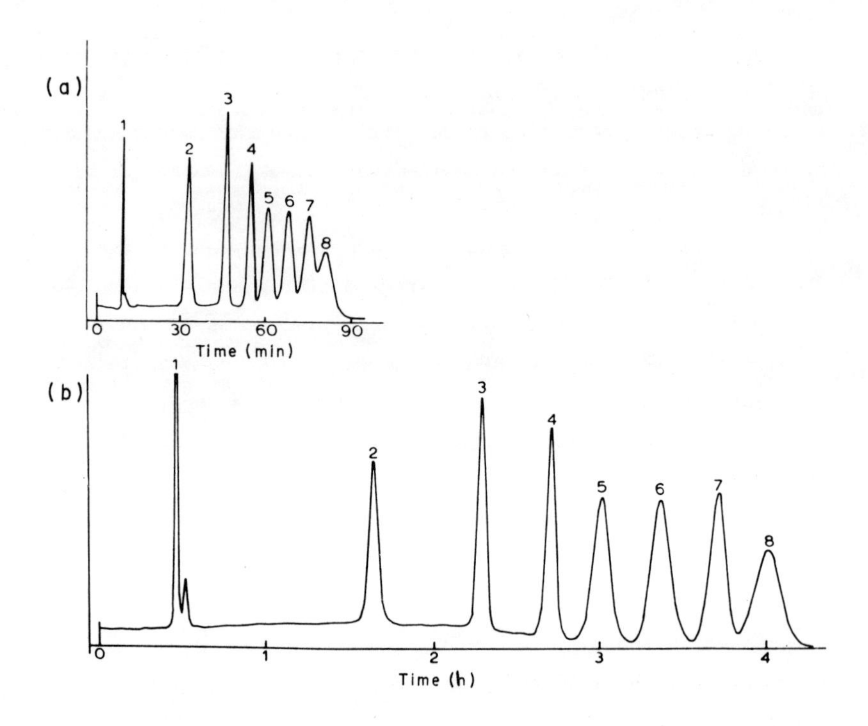

FIG. 3.5 Effect of flow rate on the separation of monosaccharides. Column: 1 m x 4 mm; Aminex A-6, 17.5 μm; mobile phase: 85% EtOH in H_2O; temperature: 65°C; flow rate: (a) 0.49 ml/min, (b) 0.16 ml/min; sample components: (1) tetramethylglucose, (2) rhamnose, (3) ribose, (4) xylose, (5) arabinose, (6) mannose, (7) glucose, (8) galactose. (From Ref. 679.)

Analysis of the carbohydrates in biological fluids are discussed in the following section.

3.6 COMPOUNDS IN BIOLOGICAL FLUIDS AND EXTRACTS

The normal and pathological constituents in biological fluids as urine, blood, and other extracts are very important but are contained in complex biological mixtures. Scott and co-workers have developed complex chromatographic systems for separating constituents of body fluids by HSIEC [118,190,562,680-707]. These systems are computerized gradient-elution HSIEC units which were used for the analysis of amino acids, indoles, organic acids, carbohydrates, and other compounds in biological fluids. Scott et al. [680] evaluated urine from leukemic and schizophrenic patients. The results were compared with the diurnal pattern of a normal male human [681] (Fig. 3.6). Human urines, blood serum, cerebrospinal fluid, and amniotic fluid were analyzed by Jolley and Scott [687]. Quantification of the resulting chromatograms revealed significant differences in the concentrations of body-fluid carbohydrates for the normal state and for pathological states such as hereditary nephritis, Lesch-Nyhan syndrome, gout, alkaptonuria, and maple syrup urine disease. In the detection of carbohydrates, the phenol-sulfuric acid method was used and the effluent was monitored at 480 and 490 nm. The body fluid, mixed with borate buffer to make anionic sugar-borate complexes, was introduced into the column via a valve [221]. Chromatograms were generated by using a borate buffer of continuously increasing concentration as the mobile phase [688,689,691,699]. A typical "carbohydrate chromatogram" contained more than 45 peaks. Ten of these (sucrose, lactose, allulose, fructose, arabinose, fucose, galactose, sorbose, xylose, and glucose) were identified also by GC of their trimethylsilyl derivatives. In the determination of UV-absorbing constituents, single- and multiple-column systems were developed and used [680-684,695-698,705].

Sequential anion- and cation-exchange columns are applicable to the determination of the UV-absorbing compounds of various body fluids, whereas parallel columns serve the comparison of body-fluid patterns from different subjects. The use of sequential or multicolumns decreases the analysis time from 24 to 14 hr. Burtis et al. [706] evaluated the phenacetin methabolic pathway and the urine from a 2-year-old girl who had

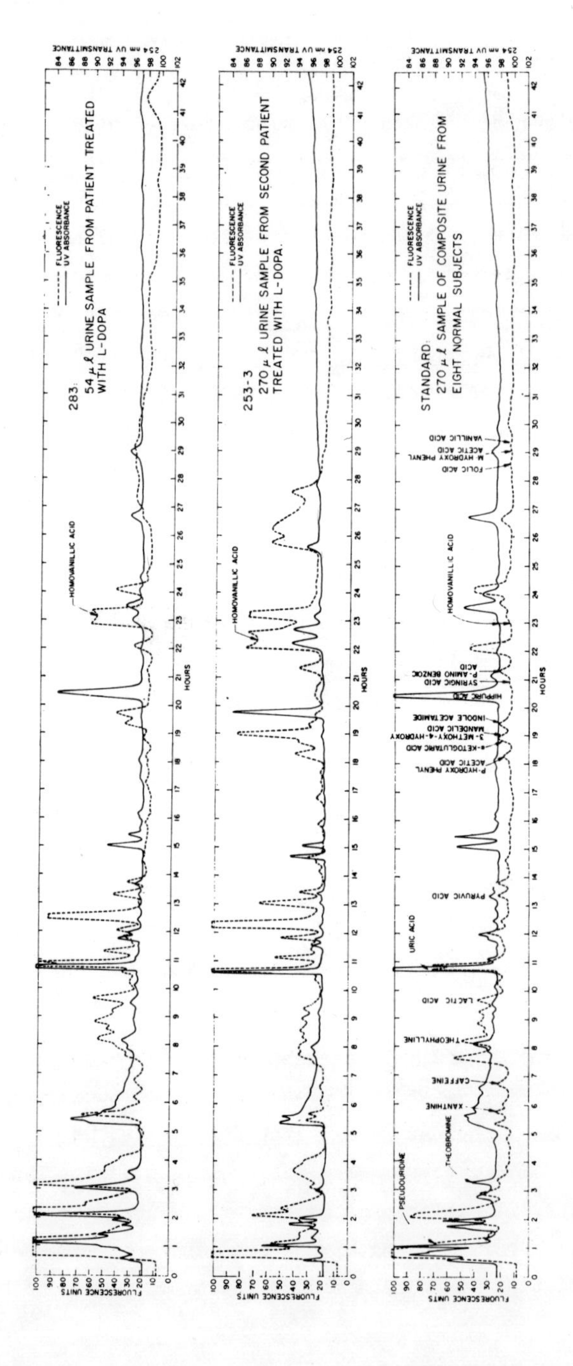

FIG. 3.6 Comparison between the separated urinary constituents of normal subjects and those treated with L-dopa. Column: 3 m x 2 mm, Pellionex SAX or Aminex A-27; mobile phase: ammonium acetate-acetic acid gradient; temperature: 60°C; flow rate: 8-12 ml/hr; detection: cerate oxidimetry. (From Ref. 592.)

a neuroblastoma. Vavich and Howell [700] separated UV-absorbing compounds in urine of normal newborns and children, using an ammonium acetate/AcOH linear gradient from 0.015 to 6.0 M (pH 4.4). Urine from a normal newborn infant and that of a composite adult were compared. Differences were found between the child and the adult in the excretion of UV-absorbing urinary components. Different amounts of creatinine, pseudouridine, uridine, hippuric acid, and ergothioneine were observed. Nanogram amounts of indoles were separated by Chilcote and Mrochek [686] who used anion- and cation-exchange columns connected in series and an ammonium acetate/AcOH buffer that was 4 M in ammonia and 5.8 M in total acetate. Scott and co-workers demonstrated the separation of more than 150 constituents from human urine. About 70% of these constituents were also qualitatively identified. The chromatograms obtained from such separations may be employed as fingerprints in evaluating body functions. To use such fingerprint profiles for diagnostic purposes, dietary effects must be considered [701-703, 708-710]. Kelley and Beardmore [708-710] studied purine and pyrimidine metabolism. They also evaluated the quantitative effect of diet on the excretion of seven purine and pyrimidine compounds (pseudouridine, uracil, 5-acetylamino-6-amino-3-methyluracil, hypoxanthine, xanthine, adenine, and uric acid). Burtis [711] and Burtis and Stevenson [712] used strongly basic anion-exchange resins and linear sodium acetate gradient elution varying in concentration from 0.015 M to 6.0 M at constant pH 4.4. The results suggest that, assuming a molar absorptivity of 10^4, nanogram quantities of some constituents can be detected.

Henderson and Jones [713] used silica-gel column and a complex gradient of $CHCl_3$ and 2-methyl-2-butanol in a pressure-forced system to determine serum organic acids (2-OH-butyric, methylmalonic, butyric, propionic, acetic, pyruvic, and lactic) in sheep exposed neutron-gamma irradiation and air blast.

Organic acids in blood cells and biological fluids were also determined by Barness et al. [714] who used a silicic acid column, a complex gradient of $CHCl_3$/tert-amyl alcohol, and glutaric and malonic acids as the internal standard in a pressure-forced system. Shargel et al. [715] determined nalidixic and hydroxynalidixic acids in human plasma and urine on a strong anion-exchange resin, using a mobile phase of 0.02 M boric acid and 0.05 M Na_2SO_4 (pH 9.0) and 600 psi. The limits of detection for the compounds were 0.25 μg of each in plasma and 2.5 μg/ml of each in urine. Metabolites of tyrosine and dopa in plasma, urine, and lumbar spinal were determined in a pressure-forced system [716]. For the separation of basic metabolites, a 13 cm x 6.3 mm Aminex A-5 column, a stepwise gradient of 0.5 N HCl and 0.75 N HCl, and a pressure of 390 psi were used. Neutral metabolites and acidic metabolites were separated on a cation-exchange column, employing a two-step (pH 3.27 and 4.47) citrate buffer gradient elution. The separated compounds were substituted mandelic acids, substituted phenylglycols, OH-substituted phenylacetic acids, vanillic acid, homovanillic alcohol, 4-OH-phenylpyruvic acid, and homogentistic acid. Aliphatic amines and polyamines in urine, animal tissues, and bacterial extracts were separated on a cation-exchange resin, using a gradient from 3.5 M sodium citrate (pH 6.32) to 3.5 M sodium citrate/ 2.35 M potassium citrate (4:1) in a pressure-forced system [717]. Among others, putrescine, cadaverine, spermidine, agmatine, spermine, and monoacetyl spermidines were determined.

Partial resolution of aflatoxins B_1, B_2, G_1, G_2, and P_1 in a synthetic mixture and B_1 and G_1 in the $CHCl_3$-extract of culture broth incubated with _Aspergillus parasiticus_ was achieved on a pellicular silica adsorbent, using $CHCl_3$/isooctane as the mobile phase at a flow rate of 1 ml/min [718]. Overlap occurred between B_1 and G_1 as well as G_2 and P_1. The practical detectability limit for B_1 was 100 ng. In reverse-phase HSLC, the order of elution is reversed; this technique might thus be preferable for the analysis of the more polar aflatoxins. The aflatoxin metabolites (B_1 and G_1) of Aspergillus flavus Link in crude extracts were separated on a small-particle silica gel column, using isopropyl ether/tetrahydrofuran (22:3) or diethyl ether/cyclohexane (3:1) as the mobile phase [497]. An MS was employed for confirmation of the peaks eluted from the column.

Propham (isopropyl carbanilate) metabolites were isolated identified from rat and goat [719]. Metabolites 17 and 18 (sulfate ester of 2-hydroxyaniline and sulfate ester of 4-OH-aniline or that of 4-OH-acetanilide) in goat urine were separated by HSLC using a Permaphase ETH polar bonded phase column and a mobile phase of EtOH/hexane (1:3) at a flow rate of 0.8 ml/min. Isolation and quantification of serum uric acid was carried out in a pressure-forced system, on a Bio-Gel P-2 column, using 0.01 M sodium phosphate buffer (pH 7.0) at a flow rate of 10 ml/hr [720].

Aldomet, a drug, was determined in blood on an Aminex A6 column, using 0.5 M ammonium acetate (pH 4.65) as the mobile phase at a flow rate of 1 ml/min at 480 psi and 60°C [369]. Rat urinary metabolites from 2-methoxy-4,6-bis(isopropyl amino) S-triazine (Prometone) were separated on a cation-exchange column at 65°C [721]. A gradient elution of 3.8 M and 5.0 M sodium citrate buffers was employed. The column was monitored by a continuous-flow liquid scintillation detector. The system was only pressure-forced.

Kirkland et al. [722] determined benomyl residues [methyl 2-benzimidazolecarbamate and 2-aminobenzimidazole] in soil and plant tissues. Other metabolites, such as 5-OH-2-benzimidazolecarbamate and methyl 4-OH-2-benzimidazolecarbamate, were determined in cow milk, urine, feces, and tissues [723]. In these separations, pellicular SCX columns and 0.025 N tetramethylammonium nitrate/0.025 N HNO_3 or 0.15 N sodium acetate/0.15 N AcOH (7:3) as the mobile phase were applied. The sensitivity is estimated at 0.1 to 0.01 ppm for these metabolites. The qualitative and quantitative analysis of ^{14}C-diphenamid (N,N-dimethyl-2,2-diphenylacetamide) metabolites was carried out on a reverse-phase column, using MeOH/H_2O (4:1) as the mobile phase [724]. The benzene extract of 3-week-old soybean plants contained 30 to 35% diphenamid, 13 to 15% N-methyl-2,2-diphenylacetamide, and 0.8 to 0.9% diphenylacetamide. The water-soluble metabolites were identified by GC.

ε-Aminocaproic acid (EACA) was determined by HSIEC in plasma, urine, cerebrospinal fluid, and erythrocyte extracts of subjects undergoing EACA therapy [725]. The method required less than 2 hr for analysis. A study of amino metabolism in the alimentary tract of pigs and the amino acid content in dry sausages was connected with separations by HSLC using a Zeocarb column [726]. The column temperature was maintained at 43°C for 103 min and switched to 75°C for the remainder of the analysis, i.e., a temperature programming was used. Free basic amino acids were determined by HSIEC in blood plasma or serum [527]. The mobile phase was 0.6 M sodium citrate buffer (pH 4.2 or 4.5). The separation could be speeded up by addition of iso-PrOH or octanoic acid to the mobile phase. A method for measuring the levels of a drug, metolazone, in urine is described by Hinsvark et al. [728,729]. A polar, brush-type packing material (Durapak Carbowax 400) and $CHCl_3$ modified with iso-PrOH were used.

Anders and Lattore [730,731] applied HSIEC to determine glucuronides and sulfate conjugates of phenol, p-nitrophenol, acetanilide, and catechol in urine. The column packed with a pellicular ion-exchange resin was maintained at 80°C. The mobile phase was 1.0 KCl in 10 mM formic acid (pH 3.0) at a flow rate of 30 ml/hr. Glucuronide conjugates of p-OH-acetanilide and 3-methoxy-4-OH-acetanilide in urine were also observed by Burtis et al. [707]. HSLC analysis of crude furocoumarin extract from Cymopterus watsonii was carried out by Stermitz and Thomas [732]. In the analysis, Corasil I column and $CHCl_3$/cyclohexane mobile phase were used. One of the unknown furocoumarins was isolated by preparative HSLC for structural investigation. Several procedures used in extracting free nucleotides from cells were compared by HSIEC [733]. The clinical use of HSLC has been recently reviewed by Scott [734]. The HSLC of other compounds, e.g., nucleic acid constituents, in biological fluids is discussed in the sections under the title of the compound family.

3.7 DRUGS AND RELATED COMPOUNDS

In GC analyses of drugs and related compounds, preparation of chemical derivatives should be made prior to the analyses because these compounds are usually nonvolatile. HSLC not only surpasses this problem but offers a method, ion exchange, with no analog in GC and which is ideally suited as a separation technique in pharmaceutical analysis for a large percentage of compounds [735]. An APC (aspirin, caffeine, and phenacetin) tablet was examined as a test mixture for cation exchange, and optimum separation of this sample was achieved with a mobile phase of distilled water buffered to pH 6.8 [596,736,737]. Components of analgesic tablets containing other compounds, e.g., salicylamide, in addition to the three mentioned above, can be easily separated by HSIEC, also [738,739]. The mobile phase should be adjusted to a pH of 9.2 with a borate buffer and the ionic strength is adjusted by the addition of 0.002 M $NaNO_3$. Increasing the ionic strength of the moving liquid reduces analysis time without significantly impairing the resolution of the sample components. It should be also noted that an anion exchanger elutes a typical analgesic in the following order: caffeine, phenacetin, and aspirin; whereas a cation exchanger gives an elution order of aspirin, caffeine, and phenacetin. Using an anion-exchange column with 0.025 M or 1.0 M tris(hydroxymethyl)aminomethane (pH 9.0) as the mobile phase, the elution order is caffeine, aspirin, and phenacetin [740,741]. Under these conditions, acetaminophen elutes after caffeine and salicylamide after phenacetin. Four to six analgesics could be separated by HSIEC using 0.1 M citric acid (pH 2.8) or 0.03 M sodium citrate (pH 3.52) as the mobile phase [742-745].

Murgia and co-workers [746,747] used ion-exchange resins and aqueous EtOH for the separation of analgesics, acetanilide and xanthine derivatives. Effects of cross linking and counterions on separation were also studied. Analgesics can also be separated on a reverse-phase column, using $CH_3CN/(NH_4)_2CO_3$ as the mobile phase in a flow-programming analysis [296]. Anion-exchange analysis of a single tablet of asthma and hay fever medication, containing ephedrine, theophilline, and phenobarbital can be carried out using 0.01 M $NaNO_3$ as the solvent [748].

Antitussives can be successfully chromatographed on a reverse-phase packing, using $CH_3CN/(NH_4)_2CO_3$ as the mobile phase [749]. Anders and Lattore [750] have managed to apply HSIEC for the separation of barbiturates, some of their metabolites, and related compounds, among others, diphenylhydantoin. Nongradient elution with KH_2PO_4 buffers has proved very suitable for separating phenobarbital, diphenylhydantoin, and their phenolic metabolites. The best conditions found for this analysis are 20 mM phosphate buffer, pH 3.5, and a column temperature of $80^{\circ}C$. Barbiturates and metabolites have been separated with a linear sodium chloride gradient at $80^{\circ}C$. Barbiturates can also be determined successfully by using an anion exchanger and 0.01

M $NaNO_3$ mobile phase [751] or using porous layer bead and 2% MeOH in heptane as the mobile phase [576].

The separation of barbiturates can also be carried out on pellicular silica gel column, using $CHCl_3$ as the mobile phase [752]. Small-particle silica gel proved also useful in the determination of the major urinary metabolite of diphenylhydantoin when using n-butyl chloride/dioxane (4:1) as the mobile phase [753]. The simultaneous measurement of diphenylhydantoin and phenobarbital in serum could be achieved by using small-particle silica gel and a mobile phase of CH_2Cl_2/MeOH/28% NH_4OH (92:7:1) [754] (Fig. 3.7).

Diphenylhydantoin may also be determined by HSLC through its conversion into diphenylketone [755]. Roos [756] used an anion-exchange column with 0.01 M sodium borate mobile phase, a cation-exchange column with 0.01 M citric acid mobile phase,

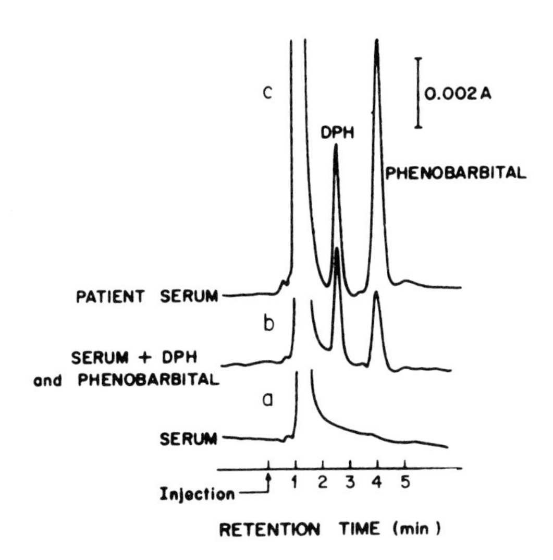

FIG. 3.7 Diphenylhydantoin and phenobarbital from human blood serum. Column: 50 cm x 2 mm, Lichrosorb silica gel, 10 μm; mobile phase: CH_2Cl_2/MeOH/28% NH_4OH (92:7:1); flow rate: 0.75 ml/min. (From Ref. 754.)

and a Permaphase ETH column with H_2O mobile phase to separate 17 barbiturates. Reverse-phase chromatography proved effective in the determination of hashish, using chemically bonded packing material and a linear gradient of 2%/min of MeOH in H_2O [662]. Reverse-phase HSLC was also applied to characterize LSD in illicit preparations [757]. The carrier was MeOH/0.1% $(NH_4)_2CO_3$ (3:2) at a flow rate of 0.65 ml/min. A combination of UV and fluorimetric detectors allowed the detection of 2-10 ng of LSD. LSD from illicit tablets can also be determined by chromatography on anion-exchange [200] or small-particle silica-gel [758] columns. Phenethylamines of forensic or toxicological interest and heroin can be determined on ion-exchange or pellicular silica-gel columns [759-760]. The phenethylamines can be injected as their HCl salts in aqueous solution. When using HSLSC, gradient elution must be employed in order to achieve a peak capacity comparable to HSIEC. A chemically bonded packing material, Permaphase ETH, gave well-separated peaks of sulfapyridine, sulfamethazone, and sulfanilamide, using n-hexane/iso-PrOH (3:2) as the mobile phase [266].

On a reverse-phase chemically bonded packing material, Bondapak C-18, sulfadiazine, sulfamerazine, and sulfamethazine can be separated, using CH_3CN/H_2O (5:95) as the mobile phase [299]. The separation of sulfamethazine, sulfaisomidine, and sulfathiazine can be carried out on a Pellidon column, using a mobile phase of iso-PrOH [311]. A mixture of sulfa drugs was separated and determined by a coupled HSLC-MS system, using Permaphase ETH and a mobile phase of heptane/iso-PrOH (7:3) [495]. HSIEC can also be well utilized in the separation of sulfa drugs as demonstrated by Kram [761] and Poet and Pu [762]. Kram [761] studied 21 sulfa drugs and used mobile phases of 0.01 M sodium borate (pH 9.2) containing sodium nitrate at varied concentrations (0.01, 0.04, 0.07, and 1.0 M). On the basis of the results, optimum $NaNO_3$ levels may be predicted for the separation of official trisulfapyrimidines. Reverse-phase HSLLC provided a highly specific and practical technique for the analysis of sulfonylurea antidiabetic agents [763]. The mobile phases were 0.01 M sodium borate containing 27.5% MeOH and 0.01 M monobasic sodium citrate containing 15% MeOH. The elution volumes for all the sulfonylureas investigated decreased as pH was raised, indicating that these compounds being weak acids are more soluble in solutions of higher pH. A polar brush-type material was employed for the analytical scale preparation of some benzodiazepines at microgram level in urine [764]. The most polar carrier used in the analysis was a 70-30% mixture of hexane and iso-PrOH, although an optimum composition was achieved at 10% iso-PrOH content. Weber [765] used the same brush-type material for the chromatography of diazepam, oxazepam, and nitrazepam, and Corasil for the separation of ketazolam and diazepam. The mobile phases were mixtures of tetrahydrofuran and isopropyl ether. Diazepam, dihydrodiazepam, oxazepam, and nitrazepam were recently separated by Rajcsanyi [766] on a new bonded small-particle polar packing material, using $CHCl_3$ or 1% iso-PrOH in $CHCl_3$ as the mobile phase.

The combination of a Zipax column and hexane/EtOH (3:1) as the mobile phase permitted low-level detection of some impurities in a 3-formylrifampin sample [767]. At the same time, it provided good measurement of the strongly polar, high-molecular weight major component. Also HSLSC may be very effective in the chromatography of synthetic organic drugs. It can be used to determine the yield obtained under various reaction conditions and to suggest preparative schemes for purifying the products of each synthesis step. 1-(3,4-Dibenzyloxyphenyl)-2-nitro-trans-prop-1-ene and 3,4-dibenzyloxybenzaldehyde were separated on a Corasil II column, using 10% $CHCl_3$ in isooctane as the moving phase [768]. Narcotics (cocaine, meperidine, procaine, and codeine) [576] and sedatives (methaqualone, chlorpromazine, and scopolamine) [769] can be separated on a Vydac pellicular silica-gel column, using $CHCl_3$/heptane/MeOH (56:42:2) as the mobile phase. The active ingredients in a cough preparation were determined by HSIEC using 0.35 N sodium nitrate (pH 5.28) as the mobile phase [770]. Mollica et al. [771] determined seven drugs which contain the imidazoline moiety (antazoline, naphazoline, phentolamine, oxymetazoline, tetrahydrozoline, and xylometazoline). They used Zipax SCX with the mobile phase of 0.01 M Na_2HPO_4 adjusted to pH 12.4. For the analysis of xylomethazoline (on a Zipax SAX column), the buffer solution was modified with 10% MeOH to increase the solubility of the drug. The separation of 6-thiopurine derivatives of pharmacological interest can be carried out by HSIEC using a gradient of KH_2PO_4 and KCl [772-774] or ammonium acetate [775]. Several of these compounds were determined as metabolites from drug-treated tumor cells.

Reverse-phase HSLC of actinomycins was reported by Rzeszotarski and Manger [776], who used CH_3CN/H_2O (1:1) as the mobile phase at a flow rate of 1 ml/min. The uniformity of the major components separated from a mixture produced by *Streptomyces parvullus* was confirmed by their chromatography in the same system, using the recycle technique of Bombaugh and Levangie [777]. Tetracycline antibiotics were separated by Butterfield et al. [778]. Griseofulvin can be separated from its related impurities in a fermentation mixture, using a Permaphase ETH column and 5% $CHCl_3$ in hexane as the mobile phase [779]. Muusze and Huber [780] presented the separation of thioridazine and its metabolites as well as the distribution coefficients of these compounds determined by a static method. The prostaglandins $PGF_{2\alpha}$ and $PGF_{2\beta}$ were separated on a silica gel impregnated with 5% $AgNO_3$, using dioxane/toluene/AcOH/2-methyl butanol (30:30:2:1) as the mobile phase at a flow rate of 1 ml/min [781]. Prostaglandins in rat kidney were determined by Dunham and Anders [782] using a small-particle silica-gel column and UV detection at 280 nm. Quantitation of plasma samples containing phenylbutazone and its metabolites can be carried out on a small-particle silica-gel column, using a mobile phase of 10% tetrahydrofuran in n-hexane [783,784]. Drug purity profiles determined by HSLC and other methods are given for 115 official drug substances by Grady et al. [785].

3.8 FOOD CONSTITUENTS

Some knowledge of the constituents present in foods may be important in several cases. On the basis of recent advances, HSLC proved to be a very useful technique for the analysis of food constituents. Aflatoxins B_1, G_2, B_1, and G_1 from a peanut butter extract could be separated on a small-particle silica-gel column, using 60% CH_2Cl_2(50% water-saturated)/40% $CHCl_3$(50% water-saturated)/0.1% MeOH as the mobile phase [579]. A method for the analysis of Clopidol, an effective coccidiostat, in poultry feed was developed by Skelly and Cornier [786]. The method involved chromatography on an alumina and on an ion-exchange column in series with 0.1% AcOH in MeOH. Food preservatives (sorbic acid and alkyl-substituted p-OH-benzoates) can be determined on Pellidon, using a mobile phase of n-hexane/AcOH (9:1) at a flow rate of 15 ml/hr [311,787]. In addition to the above compounds, biphenyl and 6-phenylphenol were also determined on a small-particle silica-gel column, using propionic acid as the mobile phase [788]. An HSIEC procedure employing an anion-exchange resin and 0.01 M sodium borate (pH 9.2) containing varied concentrations of $NaNO_3$ (0.0-0.1 M) was elaborated for the quantitation of sodium saccharide, ascorbic acid, potassium sorbate, sodium benzoate, vanillin, ethyl vanillin, and alkyl-parabens as common food additives in soft drinks [789] (Fig. 3.8).

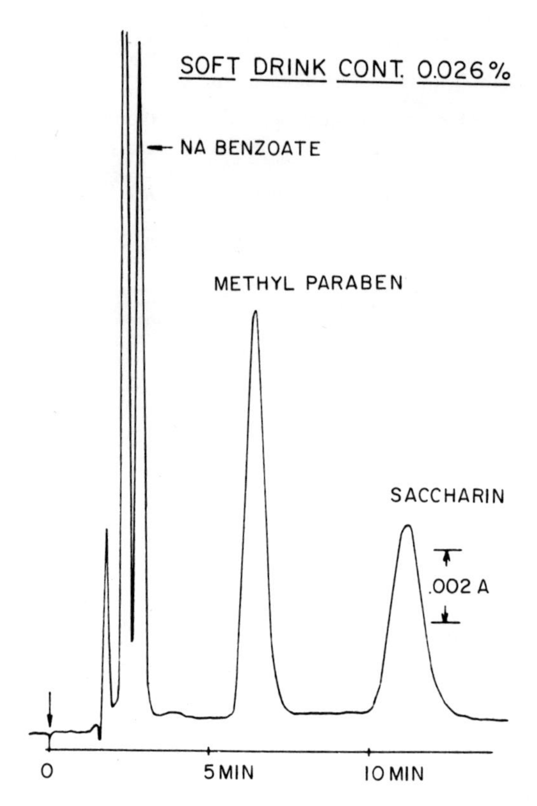

FIG. 3.8 Chromatogram of an artificially sweetened soft drink. Column: 1 m x 2.1 mm, Zipax SAX; temperature: 24°C; mobile phase: 0.01 M sodium borate and 0.03 M sodium nitrate; flow rate: 0.85 ml/min. (From Ref. 789.)

Palmer and List [790] separated microequivalent amounts of up to 15 of the organic acids commonly occurring in foods on an anion-exchange column at 70°C. The mobile phase was 1.0 N sodium formate. Stahl and Laub [791] used microcryst cellulose columns and EtOH/PrOH/H_2O (60.5:24.5:15) as the mobile phase to separate organic acids in wines and fruit juices. A comparison of GC and HSLC in the analysis of vanillin and ethyl vanillin in alcoholic solutions was presented by Martin et al. [792]. The best resolution was obtained by using HSLSC, a small-particle silica gel with $CHCl_3$/hexane (1:1). Food colors (pyrazalone, sulfanilic acid, and yellow 5) can be separated on a small-particle silica gel, using MeOH/tetrahydrofuran (1:6) as a mobile phase [645]. Quality control of flavor chemicals was carried out by HSGPC on a Poragel 60 column [793]. The mobile phase was CH_2Cl_2 at a flow rate of 100 ml/hr. A combination of HSGPC and reverse-phase HSLC was applied for the separation of complex flavor mixtures in citrus oils and rums [794]. It was possible to determine the contribution of various climatic geographic variables to the characteristics of these oils. The influence of barrel age and history and other variables on the flavor of alcoholic beverages was also studied.

Carotenoid mixtures containing also cis-trans isomers of violaxanthin as well as anthraxanthin found, e.g., in citrus peels were separated on a zinc carbonate column, using a hexane/tert-amyl alcohol gradient [795]. Carotene stereoisomers could be separated by HSLC as demonstrated by Sweeney and Marsh [796]. An HSLC procedure for the routine analysis of vitamin A, carotene, thiamine, and riboflavine in foods was elaborated by Van de Weerdhof et al. [797]. They applied aluminum oxide and silica-gel columns and an automatic injection system. A mixture of food flavoring compounds (anisaldehyde, ethyl- and methyl-vanillin, and p-OH-benzoic acid) could be separated by using a gradient elution of isooctane to $CHCl_3$ at 5%/min [138].

N-nitrosamines, the N-(2,4-dinitrophenyl) and N-nitroso-derivatives of diethylamine, pyrrolydine, and dimethylamine were determined in foods on Durapak Carbowax 400, on a small-particle silica-gel packing, and Bondapak C-18 reverse-phase columns [798]. Analysis of fried pig liver, pork luncheon meat, and the stomach contents of a dog are also described. Hobbs and Lawrence [799] determined lactose in milk and other carbohydrates by HSLC. A carbohydrate not present in milk was used as an internal standard.

Aitzetmüller [800-805] used HSGPC and HSLSC, a UV and a moving wire detector to analyze edible oils, fatty acids, and triglycerides. "Total artefact" and "polar material" peaks could be produced under appropriate conditions. Four free fatty acids in pig liver extract can be separated on a small-particle silica-gel column. The separation was completed within 15 min [806]. C-18 fatty acid esters were separated on Corasil coated with 0.8% $AgNO_3$ and 1.75% ethylene glycol, using hexane/heptane (1:1) as the mobile phase [781]. A technique for benzyl derivation of fatty acids was developed and applied for the determination of fatty acids by an HSLC/MS system [494]. The separation was carried out on a Corasil column, using a mobile phase of $CHCl_3$/heptane (1:1). From the standpoint of MS, the use of benzyl esters had no particular advantage over that of methyl esters.

HSLC is capable of resolving flavoring mixtures under mild conditions as demonstrated by Stevenson and Burtis [807], who chromatographed vanilla oil on an ion-exchange column, using a linear gradient of sodium acetate from 0.015 to 6.0 M. The taste character of beer can be studied by HSLC through the determination of hop resin compounds [808-812]. Isohumulones in beer can be separated on a Corasil column or on a column packed with small-particle silica gel, using isooctane/$CHCl_3$ (4:1) or isooctane/ethyl acetate (19:1) as the mobile phase [809]. Ion-exchange chromatography is also useful in these separations [808,813]. Palamand and Aldenhof [808] separated 22 hop resin compounds by HSLSC. The best solvent combination was isooctane/$CHCl_3$ (9:1). The contribution of the double-bond to the retention of 11 hop bitter acids was studied by Vanheertum [814] using high-speed anion-exchange separation. A linear relationship was found between this contribution and log α. Another linear relationship could be found between log t_N and the reciprocal absolute temperature [815]. A review of the application of HSLC (and classical LC) to materials of current or potential interest in brewing is presented by Siebert [816].

3.9 METAL IONS, METALLIC COMPOUNDS

Ion-exchange chromatography is a widely used method in the separation of metal ions by elution with HCl or HCl-modified mobile phases [331,817]. Hydrobromic acid mobile phase is also useful in certain cases [818]. While good separations can be obtained by classical methods, the rapid, quantitative separation of substantial amounts of such elements was not presented prior to the development of HSIEC [819]. Campbell et al. and other workers have, however, demonstrated that high-speed separation of rare earths is practical [820-824]. The technique used in the investigations consisted of loading the trivalent lanthanides and actinides in a weakly acid solution onto a cation exchanger and then eluting the metal ions with an anionic complexing agent. In the separation of elements being highly radioactive, small-particle resins should be employed to avoid radiation-induced resin damage and radiolytic gas evolution by reducing the analysis time [825]. In addition to particle size, the type of mobile phase can significantly influence the quality of separation. The importance of solvent gradient, particle size, and resin crosslinkage on resolution and efficiency has been examined by Sisson et al. [826]. Two of the most common mobile phases for the rare earths are α-hydroxyisobutyric acid and α-hydroxy-α-methylbutyric acid [820]. A ternary liquid-liquid two-phase system composed of H_2O, isooctane, and EtOH has also been used to separate metal-β-diketonates by HSLLC [827]. Veening and co-workers [828-831] demonstrated the usefulness of HSLLC in separating metals as metal carbonyl complexes, using isooctane mobile phase and UV detection. Traces of alkali metals can be determined on a cation-exchange column, using aqueous HCl as the mobile phase and a radiometric detector [832] or a thermometric detector [449]. A gradient mixture of 10 M HCl and 5 M $HClO_4$ and an anion-exchange column were used to separate 10 metal ions [833,834]. This system enabled detection of these elements via the UV absorption of their chloride complexes. Lead(II) can be separated from other metal ions on an anion-exchange column with 8 M HCl [835].

Performance of a quaternary ammonium/Zipax system for the elution of Cd(II) with HCl was studied by Horwitz and Bloomquist [836]. Metal ions, Cu(II) and Fe(III), were determined by using forced-flow LC with a coulometric detector [480]. A new method of detection based on constant potential coulometry was applied to the detection of 5×10^{-7} to 5×10^{-10} mole of metal ions separated by HSIEC [481] (Fig. 3.9). Applications of HSLSC to the separation of metallocarbones were demonstrated by Evans and co-workers [837-839]. Tris(acetylacetonato)cobalt(III) and tris(acetylacetonato)chromium(III) could be eluted with CCl_4, toluene or p-xylene from a gel column [840]. Seven different metal ions (Cd, Zn, Fe, Pb, Cu, Co, and Mn) could be separated from each other by using forced-flow IEC with solutions of HCl in iso-PrOH/H_2O or acetone/H_2O as the mobile phase [841].

3.10 NUCLEIC ACID CONSTITUENTS

The constituents of nucleic acids are perhaps the most studied field of compounds in separations by HSLC. This section of the book demonstrates application of HSLC instrumentation to these investigations, but only to a limited extent in order not to overstress this field at the expense of others. It is preferred to refer to more extensive and remarkable surveys [570,842-844]. All the chemical characteristics of the nucleic acid constituents

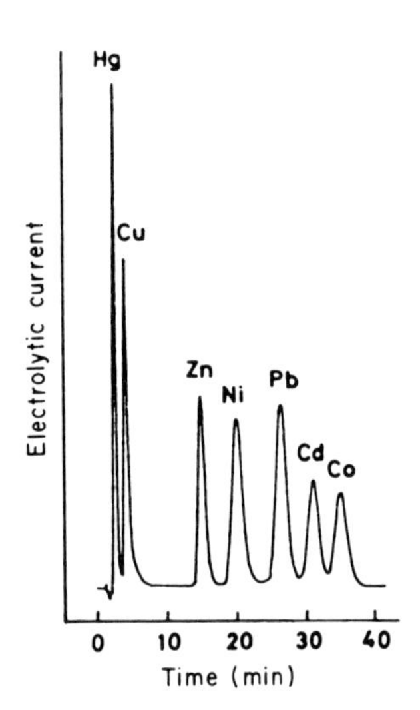

FIG. 3.9 Heavy metal ions separation. Column: 9 cm x 9 mm, Hitachi No. 2611; mobile phase: 0.15 M sodium tartarate, 0.09 M NaCl, pH 3.5; flow rate: 2 ml/min. (From Ref. 481.)

suggest that HSIEC is a very functional and desirable mode of separation, although the degree of ionization is subject to temperature, pH, and ionic solvent. In the case of bases and nucleosides, strongly acidic cation exchangers whereas for nucleotides, strongly basic anion exchangers are preferred. To incorporate fast mass transfer, resins with small diameter should be chosen. (See Fig. 3.10.) To determine which parameter to adjust for optimum results in separations of nucleic acid constituents, one should keep in mind several useful guidelines described by Horvath [264,845], Uziel et al. [846], and Burtis [847] as well as Hanson [848]. The pH of the eluent dramatically effects both retention and dispersion of solute bands. For rapid separation of bases and nucleosides, the pH range from 5.2 to 5.5 was found most adequate. Increasing or decreasing ionic strength by 0.1 M or less is usually sufficient, therefore gradient elution may be extremely useful. Temperature effects separations

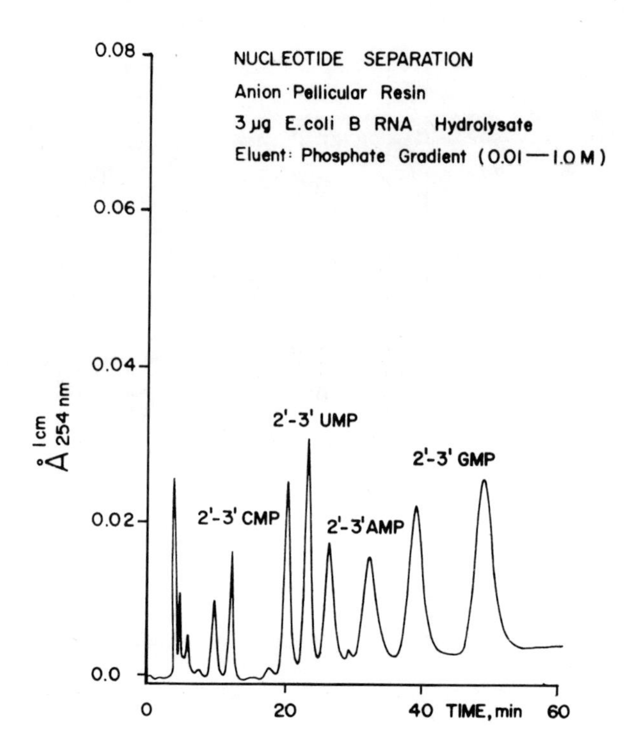

FIG. 3.10 Chromatogram of E. coli hydrolyzate. Column: 3 m x 1 mm, Pellionex SAX; mobile phase: gradient of KH_2PO_4 from 0.01 M to 1.0 M; temperature: 80°C; flow rate: 6 or 12 ml/hr. (From Ref. 842.)

in two ways. Increasing the rate of diffusion and the various kinetic processes with temperature, sharper and more symmetrical peaks can be obtained. On the other hand, retention volumes usually decrease, while relative retention values often increase with the decreasing temperature. High pH appears to be more effective than high temperature to obtain high resolution. The greatness of the flow rate to be used is limited because at high flow velocities peak spreading is governed by mass transfer in the stationary phase.

In HSIEC separations of nucleic acid constituents, pellicular ion exchangers developed by Horvath et al. [264,269] are very frequently, almost solely, used. Such type anion exchangers have been used for the separation of nucleotides, with dilute KH_2PO_4 solvents and KCl gradients [772-774,849-861]. 3,5-cyclic Adenosine monophosphate can also be determined by anion-exchange chromatography with HCl eluent [862-866]. The excellent resolution of other anion exchangers has been illustrated by several investigators, who separated nucleotides or purine-base modified nucleotides [867-871]. Pellicular and conventional ion exchangers were compared by Breter and Zahn [872] in the separation of the four major deoxyribonucleotides and d-Ino. They used a linear gradient of ammonium formate. A disadvantage of using conventional resins was that a column had to be prepared each day. Fourteen nucleotides could be separated on a pellicular anion-exchange column by using a gradient of $NH_4H_2PO_4$ from 0.01 M (pH 3.0) to 0.125 M (pH 4.3) [873]. Horvath and Lipsky [378] used a liquid ion exchanger and KH_2PO_4 as the mobile phase to separate nucleotides. Nucleotides in cell [772,870,874] and tissue [858, 859,871,875,876] extracts as well as other biological fluids [856,877-882] could also be determined by HSIEC.

Impurities in nicotinamide adenine dinucleotide can be determined by anion exchange, using KH_2PO_4 gradient from 0.002 M to 0.5 M [856]. For the determination of nucleosides and bases, cation exchangers with ammonium formate and potassium or ammonium phosphate mobile phases have been used [621,845,883-890]. In certain cases, ammonium or sodium acetate, $NaNO_3$ or HNO_3 may be employed [596,853,891-894]. In some instances, anion exchange chromatography has been used to separate nucleosides and bases [895,896]. Kelemen and Degens [381] used a gradient system of 1 N NaCl in NH_4OH and NH_4Cl to separate the five major nucleic acid bases. The level of dCyd in urine was determined by HSIEC using 0.3 M citric acid/sodium citrate buffer (pH 5.0) [897]. Human plasma levels of the endogenous oxypurines, Xan and Hyp, could be measured by HSLSC using a silica-gel column and a solvent system of ethyl ether/PrOH/5% AcOH in H_2O (35:14:4) flowing at 45 ml/hr [898]. Nucleosides and bases can also be determined in other biological samples [545,899,900].

Weak anion exchangers have been used by Gabriel and coworkers [901-903] to separate oligonucleotides. A gradient consisting of 40 ml each of 0.001 and 1.0 M of ammonium or potassium sulfate, all buffered at pH 4.4 with 0.001 M ammonium acetate, was used as the mobile phase. (See Fig. 3.11.) Vlasov et al. [904] used DEAE-cellulose or Dowex-1 columns in a pressure-forced system to separate oligonucleotides. Egan [905] separated oligonucleotides on a reverse-phase column, using gradient elution of ammonium acetate/AcOH from an acetate concentration of 0.5 M to 1.8 M. Duch et al. [906] elaborated an HSIEC method which allowed identification of both termini in an oligonucleotide of any length. They used anion-exchange resins and an exponential gradient of KH_2PO_4 from 0.005 M (pH 3.0) to 1.0 M (pH 4.4). Kelmers and Heatherly [907] separated tracer-labeled aminoacyl-tRNA's on a reverse-phase column in an HSLC (max 500 psi) system. The time required to complete an experiment could be shortened to 30 min.

3.11 PESTICIDES

Although pesticides can be analyzed by GC, these compounds frequently require preliminary time-consuming cleanup steps and the synthesis of derivatives for analysis. Therefore, HSLC may be a complementary method for the analysis of pesticides. Substituted urea herbicides (linuron, diuron, monuron, and fenuron) were chromatographed by HSLLC on BOP stationary phase, using several types of liquid as mobile phase [88,636,908]. Fenuron, monuron, and diuron could also be separated on a reverse-phase column, using $MeOH/H_2O$ (1:1) as the mobile phase [299]. The

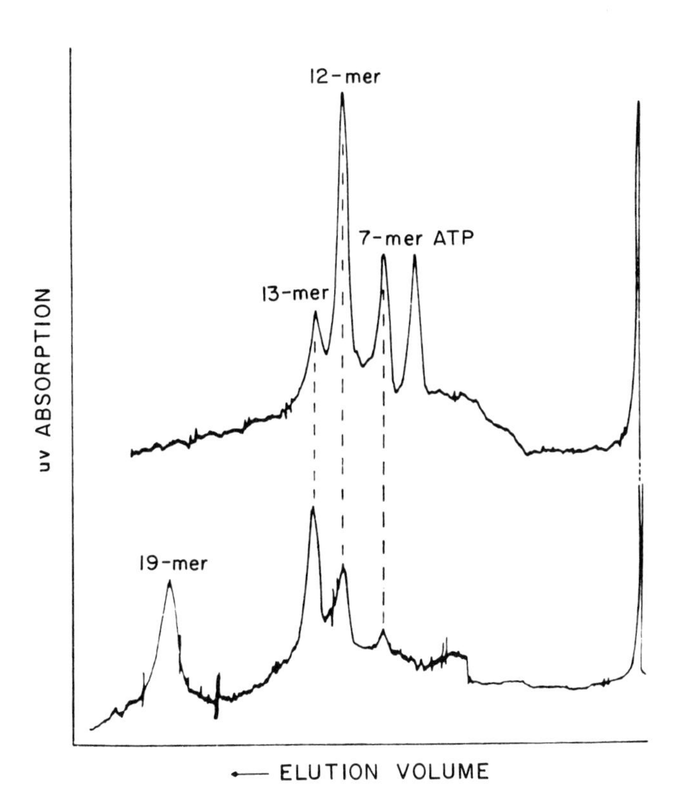

FIG. 3.11 Oligonucleotide separation. Column: 50 cm x 1.2 mm, Pellionex AL-WAX; temperature: 50°C; mobile phase: gradient of ammonium sulfate, 0.001 to 1.0 M buffered to pH 4.4 with 0.001 M ammonium acetate; flow rate: 1 ml/min. (From Ref. 903.)

separation of a mixture of thiohydroxamates was reported by Kirkland and De Stefano [294], using Permaphase ETH and a mobile phase of 10% $CHCl_3$ in hexane. The use of 10% dioxane in iso-octane as the mobile phase improved and speeded up separation [254].

A high-sensitivity detection of diuron (down to 5 ng) has been achieved on a polar chemically bonded phase with hexane modified by 3% dioxane carrier liquid [660]. Good separations of the N-methyl carbamates as dansyl derivatives of the phenyl moiety and the methylamine were achieved on Corasil in HSLSC and also on Zipax coated with 0.5% BOP [454]. The mobile phases were 2% acetone in hexane and 5% MeOH in hexane, respectively. Insect juvenile hormone and synthetic intermediates were separated on a Corasil column, using 0.5% ethyl ether in hexane as the mobile phase [909].

For the analysis of esters of phenoxyacetic acid-base herbicides, Permaphase ODS packing material was used with an H_2O/MeOH (2:3) mobile phase [908]. The free acids eluted with the solvent front under these conditions can be chromatographed by reducing the MeOH content in the mobile phase and adding dilute phosphoric acid. The separation of a mixture of thermally unstable pyrethins can be achieved rapidly and quantitatively on a silica-gel adsorbent, using 20% diethyl ether in hexane as the mobile phase [910]. Quantitative studies of pyrethins were carried out also on reverse-phase chemically bonded packing material and by HSGPC [509,662,911]. Insecticide carbaryl can be separated from its primary decomposition product 1-naphtol on a BOP or TMG column, using n-heptane saturated with the stationary phases BOP and TMG, respectively [575,912]. A baseline separation of a synthetic mixture containing the active ingredient of Lannate methomyl insecticide could be obtained on a 1% BOP column with hexane modified by 7% $CHCl_3$ [325]. Many of the chlorinated hydrocarbon pesticides were also chromatographed on a BOP column with heptane/$CHCl_3$ or on a reverse-phase column with MeOH/H_2O mobile phases [913]. Lindane, Endrin, Heptachlor, DDT, and DDD may be separated on a reverse-phase packing material with MeOH/H_2O (7:3) [914], on a BOP column with isooctane [57, 319,518], or on a pellicular silica column with hexane [198, 398]. Methoxychlor and other pesticides can be chromatographed on a BOP column with isooctane [239] or hexane [915]. (See Fig. 3.12.)

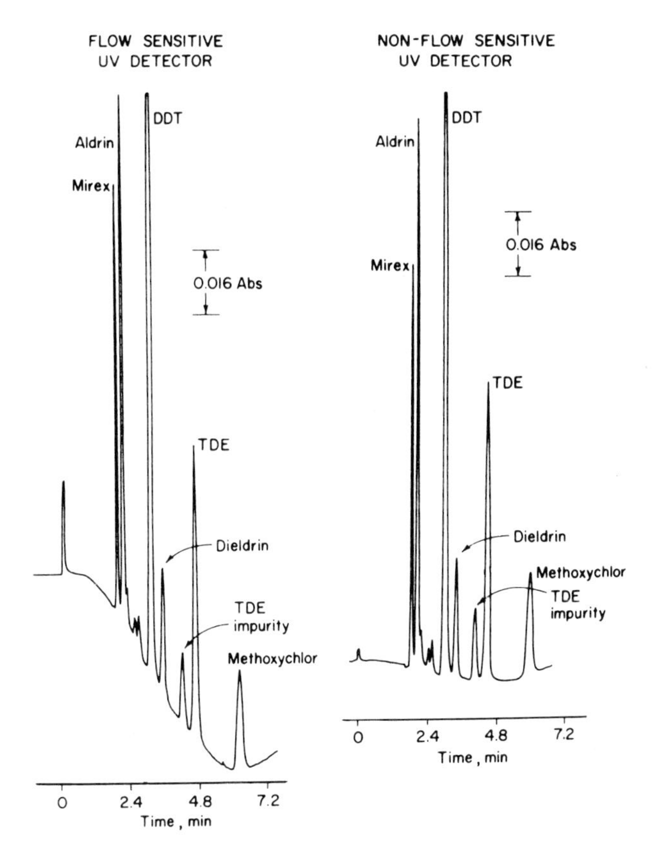

FIG. 3.12 Flow-programmed separation of insecticides. Column: 33% BOP on Micropak silica gel, 10 μm, 50 cm x 2.4 mm; mobile phase: isooctane; flow rate: 30 ml/hr to 100 ml/hr at 10%/min. (From Ref. 319.)

The potential use of HSLC in residue analysis (dithianone, DDT, and 2,4-dichlorophenoxyacetic acid) was studied by Eisenbeiss and Sieper [916], using a pellicular silica-gel column and a mobile phase of heptane/ethyl acetate (96.5:3.5). Parathion, Folpet, and Imidan were determined on a polar chemically bonded stationary phase, using isooctane solvent [917]. Pesticide residue analysis of Dyfonate can be carried out by HSLSC with a 2.5% $CHCl_3$ in isooctane mobile phase [918]. Hexachlorophene can be chromatographed on a small-particle silica gel, using a mobile phase of hexane/CH_2Cl_2/iso-PrOH/AcOH (89:8:1:2) [919] or that of hexane/n-butylchloride (55:45) [920]. An aluminum oxide column and a mobile phase of benzene/diethyl ether were used to separate Parathion and methyl-Parathion [630]. HSLLC was used to detect and isolate impurities in pesticide Abate. The chromatographic procedure was combined with a bioassay experiment [266,660,921]. Diazinon and its metabolites were separated by Machin using HSLC [922].

3.12 PHENOLS AND RELATED COMPOUNDS

Sleight [627] has studied not only the electronic effects of substituents on the retention of anilines but also those of phenols. The results suggest that electron-withdrawing substituents increase phenol retention and vice versa. Retention is decreased by introducing ortho substituents into phenols, and for ortho alkyl phenols this decrease appears to be in correlation with the size of the alkyl group. To directly extrapolate separations by TLC to HSLSC, three hindered phenolic antioxidants were separated on a silica-gel plate and on a porous silica layer bead column [631]. In both cases the mobile phase was hexane with 1% iso-PrOH modifier. The elution order was the same on TLC and by HSLSC. Also on silica gel, phenol, xylenol, and p-nitrophenol were chromatographed with CH_2Cl_2 moving liquid [923]. The 2,6-, 2,5-, and 3,4-xylenols were separated on a column filled with Mercosorb SI 60 fine silica particles, using CH_2Cl_2 as the mobile phase [924].

The mononitrophenols can be separated on a Pellidon column, using cyclohexane/AcOH (9:1) as the mobile phase [925]. Phenol and 4-isopropylphenol were separated on a surface etched glass bead column coated with 0.2% BOP, using n-heptane as the mobile phase [315]. Phenols and naphtols could be separated on a silica-gel column deactivated with H_2O, using hexane as the mobile phase [97]. A macroporous polyacrylate resin, XAD-7, was used in conjunction with $MeOH/H_2O$ or basic aqueous eluents for the chromatography of 21 phenols [926]. A simple linear relation was found between log K and MeOH-concentration in the mobile phase. Separations of several phenol derivatives could be carried out by using the Fractonitril stationary phase [927]. Huber [255] separated dimethyl- and trimethylphenols as well as 3-methylphenol on a porous material coated with 6% Fractonitril, using isooctane as the mobile phase. Trace phenol and isomeric hydroxybenzoic acids could be determined in aqueous solutions by means of aqueous HSIEC [928,929] (Fig. 3.13).

3.13 POLYMER RESINS

As the physical properties of polymers of relatively high-molecular weight are strongly related to their molecular weight, the separation of polymer compounds can be achieved by a technique, called steric exclusion or gel permeation chromatography, based on molecular size [525,527,930-939]. The degree of retention depends on the size of the sample component molecule relative to the size of the pores through which the selection diffusion of solute molecules proceeds. Therefore, molecules larger than the pore size may possibly be completely excluded from the pores and will elute from the column before the smaller molecules. This technique of separation is preferred at molecular weight ranges above 2000 units. From the results obtained by GPC, one can determine the polymer molecular weight distribution or an average molecular weight. Polyvinyl alcohols from various methods of production were separated on four columns in series, each packed with gel of different pore size [243].

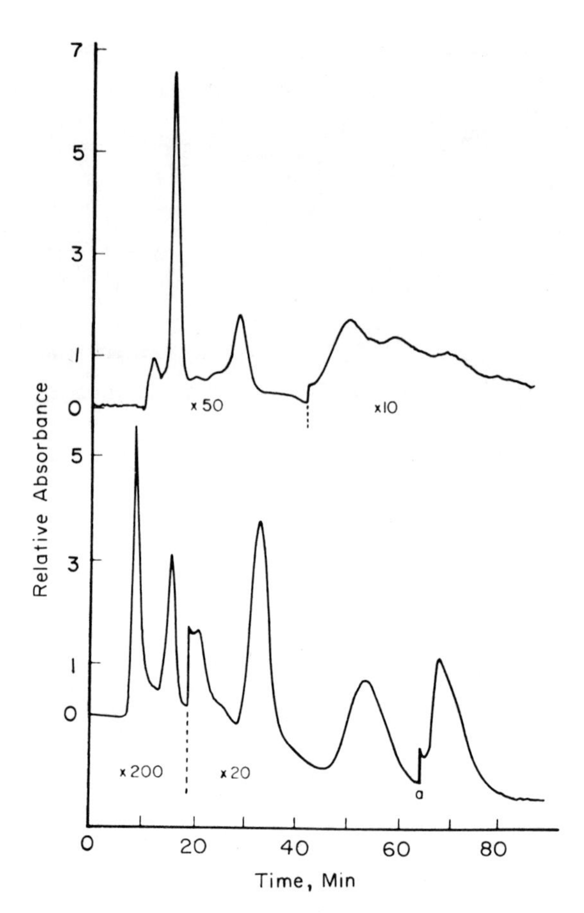

FIG. 3.13 Analysis of industrial waste waters contaminated with phenols from coking operations. Column: 2 m x 2.1 mm, Bondapak C-18; mobile phase: MeOH/H_2O (1:4); temperature: 46°C; flow rate: 1 ml/min. (From Ref. 929.)

The separation of surfactants by HSGPC can be carried out on gels with pores of 500 Å, using tetrahydrofuran as the mobile phase [940]. The separation of crude oil before and after distillation has shown the potentiality of HSGPC as a distillation analyzer and to replace or supplement distillation and GC because the method is nondestructive and usable at ambient temperature. Aqueous steric exclusion analysis of polysodium silicate was carried out on a porous silica column [941]. On a column packed with 5-6 μm porous silica microspheres with pores of 350 Å, polystyrene fractions with molecular weight of 2000, 51,000, and 411,000 were separated, using tetrahydrofuran as the carrier [60]. A mixture of low and high molecular weight polystyrene polymers can be separated also on glass columns [942].

The monomers and polymers in phenol formaldehyde resins can be clearly identified by HSGPC [943]. Ede [944] developed an HSGPC method for determining the molecular weight distribution of Nylon 6. Using this method, good agreement was obtained by Ede between number average molecular weights determined by HSGPC and by end group determinations. The variation of distribution with polymerization time was also studied. Although as described and noted above, GPC is normally used for large molecules with molecular weight above 2000 units, small molecules may also be rapidly separated by this method [940,945, 946]. Polyethylene glycols could be chromatographed successfully on a Sephadex LH-20 column, using H_2O as the mobile phase (0.8 ml/min), and detected with a Christiansen-effect detector [405].

A series of nonylphenolethylene oxide adducts varying in chain length from 1 to 20 ethylene oxide units was chromatographed by Huber et al. [947]. A size separation of isocyanate oligomers (from monomer to tetramer) was carried out on a column filled with Biobeads SX-2, using $CHCl_3$ as the mobile phase [660].

3.14 STEROIDS

Representative members of several classes of steroids were separated by Siggia and Dishman using HSLC [948]. Reverse-phase chromatography was used throughout their study because the relatively polar steroids investigated were quite polar and eluted slowly enough using nonpolar eluents. Therefore, the mobile phase consisted of a mixture of H_2O/MeOH or H_2O adjusted to pH 11.5 with NaOH. The retention time of the fast eluting, more polar adrenal corticosteroids decreased with increased stationary phase loading while the less polar steroids showed a minimum retention at a given loading and then increased retention at higher loadings. This behavior can be explained by the fact that as the loading is decreased the contribution of the liquid-solid adsorption is increased and when liquid-liquid partition becomes predominant, the retention times increase with loading. For the polar steroids, the adsorption process was predominant over the loading range (18-35%) studied by Siggia and Dishman. For the separation of nine corticosteroids, stepwise flow programming was employed. Under the same conditions, the progestins eluted after all the corticosteroids and androgens chromatographed. Therefore, to determine these compounds, an H_2O/MeOH mixture was recommended as mobile phase. Progesterone was, however, not satisfactorily separated from 4-pregnene-20β-ol-3-one. Huber and co-workers [949-952] carried out the quantitative analysis of trace amounts of estrogenic and corticoid steroids in blood and pregnancy urine. Two coexistent phases of the ternary system H_2O/MeOH/isooctane were used as the stationary and mobile phases. Partition coefficients of 28 steroids in this ternary system were determined earlier [19,21]. Henry et al. [953] employed normal phase, reverse-phase, and ion-exchange chromatography to separate 2,4-dinitrophenylhydrazine derivatives of all major types of steroids and the use of both UV and RI detectors for steroids is discussed. Conventional liquid-liquid partition chromatography proved to be the most useful for separating mixtures of closely related steroids. Ethyleneglycol, BOP, as well as cyanoethylsilicone were used as stationary phases. Progesterone represented about the least polar steroid that could be partitioned on a BOP column. For the separation of estradiol glucosiduronic acid from impurities, a strong anion exchanger was applied with a linear gradient of $NaClO_4$ in phosphate buffer. Estrogenic steroids in pregnancy and nonpregnancy urines were analyzed by HSLSC using a Corasil column with an EtOH/hexane eluent at flow rates between 0.5 and 3.0 ml/min [954]. Detection limits were approximately 20 ng. Nonpregnancy urine samples analyzed contained significant amounts of estrone and estradiol but no detectable estriol. A mobile phase of hexane/$CHCl_3$/MeOH (75:22:3) proved also useful in the separation of estrone, estradiol, and estriol on a pellicular silica-gel column [576].

The above separation could also be carried out on a reverse-phase column, using CH_3CN/H_2O (3:7) as the mobile phase [955]. Six partition systems, using BOP, cyanoethylsilicone, ethyleneglycol, a mixture of BOP and ethyleneglycol, Permaphase ETH, and Permaphase ODS were investigated in the separation of equine estrogens and equine estrogen conjugates [956]. Permaphase ETH chemically bonded phase provided the most useful separation of the equine estrogens, using 2% tetrahydrofuran in hexane as the mobile phase. A three-component two-phase system composed of EtOH/CH_2Cl_2/H_2O was used by Hesse and Hövermann [957] to separate estrogens and corticosteroids. Touchstone and co-workers [958,959] separated and quantitated cortisol, cortisone, and aldosterone on a small-particle silica-gel column with a mobile phase of $CHCl_3$/dioxane (20:1). The same column was used for the quantitative determination of corticosteroids from plasma. Eight natural corticosteroids could be separated on a water-deactivated silica-gel column, using a gradient elution of MeOH/$CHCl_3$ or diethyl ether/hexane and a combination of UV and a moving wire detector [960-962]. Schulten and Beckey determined steroids in rat serum, on a Spherosil XOA 400 column with a ternary system of EtOH/CH_2Cl_2/H_2O. The separation of corticosterone, cortisone, and hydrocortisone could be carried out on a silochrom column with a mixture of $CHCl_3$/EtOH (97:3) as the mobile phase [95] or on a pellicular silica column with hexane/$CHCl_3$/MeOH (60:38:2) [963]. Hydrocortisone, hydrocortisone acetate, estrone, and methyl testosterone can be separated on a chemically bonded reverse-phase packing, using CH_3CN/H_2O (1:4) as the mobile phase [964]. Progesterone, 19-nortestosterone, estrone, corticosterone, and cortisone alcohol can be chromatographed successfully on a Permaphase ETH column, using a linear gradient of hexane to hexane/iso-PrOH (4:1) at 2%/min [266] (Fig. 3.14).

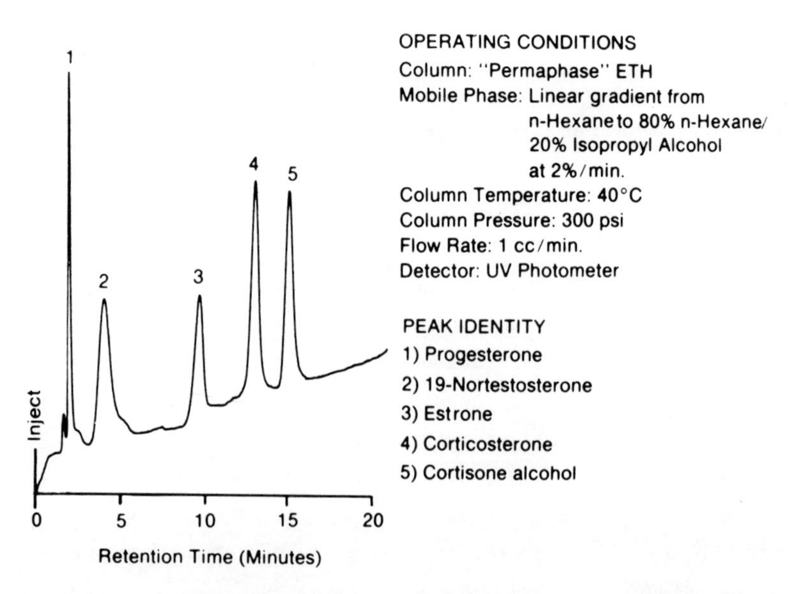

FIG. 3.14 Separation of steroids. Column: Permaphase ETH; mobile phase: linear gradient from n-hexane to 80% n-hexane/20% isopropyl alcohol at 2%/min; temperature: 40°C; pressure: 300 psi; flow rate: 1 cm^3/min; detector: UV photometer. Peak identity: (1) progesterone; (2) 19-nortestosterone; (3) estrone; (4) corticosterone; (5) cortisone alcohol. (From Ref. 266.)

HSLC assay methods were established for six major corticosteroids and two androgenic steroids in their common dosage forms [965]. Two solvent systems, $MeOH/H_2O$ and CH_3CN/H_2O, and a Bondapak C-18 reverse-phase column were used. Methyl prednisolone and methyl prednisolone-21-acetate can be separated on a reverse-phase packing, using $MeOH/H_2O$ (1:9) as the mobile phase at a flow rate of 1 ml/min [265]. Dexamethasone and prednisolone were separated by HSLSC using a microporous silica-gel column and a gradient starting with $CHCl_3$ and ending with 3% MeOH in $CHCl_3$ [966]. The presence of the methyl group at position 16 had a steric influence on the adsorption ability of the 17-OH group. A rapid separation of eight human steroids could be obtained with a complex gradient of 2% iso-PrOH in hexane and CH_2Cl_2/iso-PrOH (1:1) on a chemically bonded polar packing material [967]. Six testosterone and estradiol derivatives could be separated on a reverse-phase column, using CH_3CN/H_2O (2:3) as the mobile phase [299].

The more common 17-ketosteroids could be detected as their DNPH-derivatives from urine and plasma [968,969]. Pellicular packing materials coated with BOP and isooctane as the mobile phase were employed. Several hydroxy-containing steroids were derived and separated as their benzoates by Fitzpatrick and Siggia [969,970]. Mixtures of MeOH and H_2O as the mobile phase as well as a reverse-phase packing material were used in the separation. The resolution of the benzoates was 1-5 times greater than the DNPH-derivatives for the same analysis time. Digitoxigenin and its glycosides were chromatographed by Evans [971] on ion-exchange columns with distilled water modified by alcohols. On a reverse-phase column, ecdysone, cyasterone, and ponasterone-A could be separated, using $MeOH/H_2O$ (7:3) as the mobile phase at a flow rate of 3.1 ml/min [972]. Kram [973] suggested the use of BOP as a suitable stationary phase for separation of equine estrogens and indicated that variations in the composition of the mobile phase produced minor changes in the relative retention times of some steroids. Permaphase ODS reverse-phase packing material was used by Bailey and Brittain [779,974] to separate steroid fermentation mixtures as well as fluocinolone acetonide and fluocinolone acetonide acetate. Fitzpatrick [975] studied chromatography of 11 steroids on CN-terminated stationary phases. Particular selectivity toward the separation of corticosteroids with an α,β unsaturated keto group was noted. Fluocionide ointment preparations and other steroids were chromatographed on chemically bonded polar phases, using a mobile phase of isooctane/iso-PrOH/CH_3CN (69:18.5:12.5) [976]. Mollica and Strusz [977] determined corticosteroids in ointments and creams. Karger and Berry [978] separated steroids in body fluids and extracts. A column loaded heavily with formamide proved effective in the separation of male sexual steroids, using CH_2Cl_2 as the mobile phase [97].

3.15 VITAMINS

HSLC presents a well-suited method for the analysis of fat- and water-soluble vitamins [979-985]. Fat-soluble vitamins are best chromatographed by the reverse-phase technique whereas the ion-exchange separation process can be well applied to the determination of water-soluble vitamins. When a multiple vitamin sample is being separated it may be necessary to change the mobile phase by gradient elution because these vitamins vary widely in their structures and chromatographic properties. A typical gradient elution separation of six fat-soluble vitamins can be carried out on a chemically bonded reverse-phase packing material, using gradient from H_2O to MeOH at 5%/min. Also on this column, a tocopherol mixture may be separated using a gradient from MeOH/H_2O (4:1) to MeOH/H_2O (17:3) at 1%/min. As vitamin K is the most water soluble of the fat-soluble vitamins, it is chromatographed in a solvent of $MeOH/H_2O$ (1:4). The water-soluble vitamins to be chromatographed may be divided into two classes: those that separate by anion exchange and those that separate by cation exchange. The vitamins B_2, B_3, and B_6 can be determined by cation exchange at pH 4.2 in KH_2PO_4 buffer modified by sodium perchlorate. The separation of the vitamins B_1 and B_{12} requires a pH of 9.2. Ascorbic acid, folic acid, and niacin can be chromatographed on an anion exchanger with a mobile phase adjusted to pH 7.0 with various concentrations of ionic modifier added. A mobile phase of 0.02 M sodium acetate at pH 4.6 is also useful [743].

HSLSC can also be employed for the analysis of vitamins. On a pellicular column, A, D_2, and E vitamins were separated using $CHCl_3$ as the mobile phase [986]. A mixture of vitamin A palmitate, vitamin A acetate, vitamin D_2 was separated on a Pellumina column with 1-chlorobutane mobile phase [632] or on a Corasil column with 10% $CHCl_3$ in isooctane [987]. A mixture of vitamins A_1 and D_2 were chromatographed on a pellicular column, using $CHCl_3$/isooctane (3:1) as the mobile phase [981]. A number of fat-soluble vitamins were successfully chromatographed on a column packed with a small-particle silica-gel packing, using 1% iso-PrOH in hexane as the mobile phase [982]. A mixture of hexane/CH_2Cl_2/iso-PrOH as the mobile phase was used for the analysis of vitamin A acetate and vitamin E on a silica-gel adsorbent [355]. Vitamin B_3 and other compounds could be determined in breakfast cereal extracts on cation-exchange columns [200,985]. HSLC could be used in the manufacture of vitamins to monitor raw materials and to control quality in final product. A therapeutic vitamin-mineral tablet was analyzed by HSLC using a Corasil column and $CHCl_3$/isooctane (3:1) as the mobile phase [988].

Tocopherol acetate, ergocalciferol, and retinol can be separated on a small-particle silica-gel column (30 cm x 2.7 mm), using hexane/$CHCl_3$ as the mobile phase [661]. Tocopherols could be separated in a pressure-forced system employing a hydroxyalkoxypropyl Sephadex column and fluorescence detection [989] or by HSLSC using a Corasil column with diisopropyl ether/hexane (5:95) as the mobile phase [990]. Results obtained by GC and HSLLC, using a BOP column with hexane as the mobile phase, in vitamin A analysis have been given by Vecchi et al. [991]. The analyses can be carried out in a shorter time by using small-particle silica gel with an average particle diameter of 5 μm (Fig. 3.15).

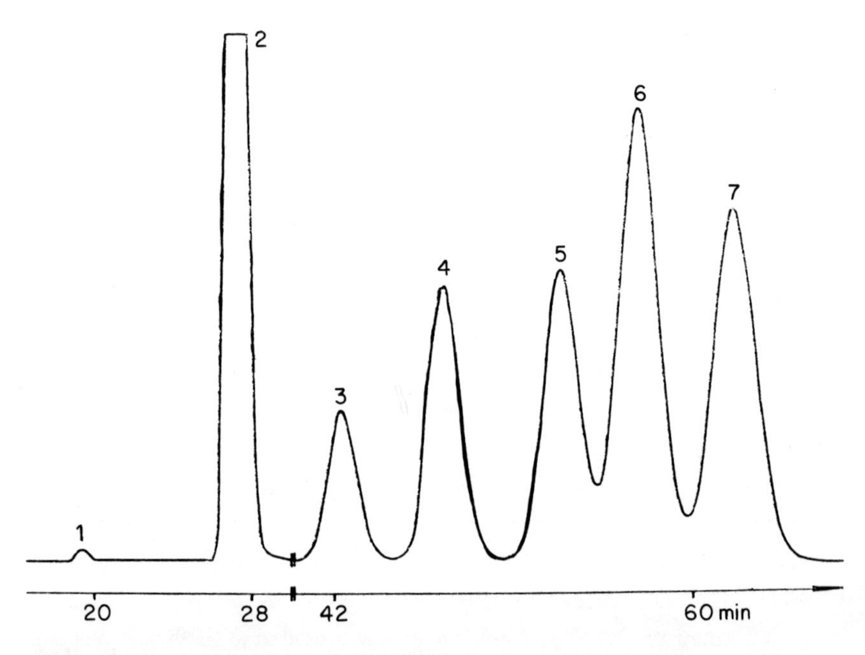

FIG. 3.15 Separation of the vitamin A acetate isomers. Column: 3 m x 2 mm, Corasil coated with BOP; mobile phase: hexane; sample components: (1) anhydro-vitamin A, (2) naphthalene (internal standard), (3) 11,13-di-cis-vitamin A acetate, (4) 11-cis-vitamin A acetate, (5) 13-cis-vitamin A acetate, (6) 9-cis-vitamin A acetate, (7) all-trans-vitamin A acetate. (From Ref. 991.)

3.16 MISCELLANEOUS

The separation of a mixture of di-n-alkyl phthalates was carried out on a pellicular silica-gel column coated with 0.8% BOP, using hexane as moving solvent [320]. Separation of diphenyl-phthalate from insecticide EPN was achieved on a 0.5% BOP column with hexane [88]. Didecyl-, dibenzyl-, and decylbenzyl phthalate plasticizers, which can be determined by GC only after being saponified, have been chromatographed successfully on a pellicular silica-gel column coated with BOP, using isooctane carrier liquid [631] or on a small-particle silica-gel column with 10% CH_2Cl_2 in hexane [239].

Separation of the three isomers of phthalic acid was carried out on a silica-gel column, by simple HSLSC [992]. Phthalic acid isomers make a suitable test mixture for the evaluation of a strong cation-exchange column performance using H_2O as the mobile phase [993]. Near the pK of the compound, pH has a great effect on retention and resolution and this effect may cause a change in the elution order. By adjusting the pH to 2.75, the elution order is as follows: phthalic, terephthalic, and isophthalic acids. The retention times of the compounds are strongly influenced by ionic strength changes. A mixture of phthalate plasticizers (di-iso-decyl-, di-iso-octyl-, and dibutyl-phthalate) could be separated from the bulk polymer, using a Corasil column and a mobile phase of isooctane/butyl acetate (98:2) at a flow rate of 0.6 ml/min [994,995]. The separation of four alkyl phthalates could be carried out in less than 30 sec on a 50-cm Zipax column coated with 0.8% BOP [996]. Phthalic acid isomers can be separated on an ion-exchange column, using 0.02 M $NaNO_3$/0.01 M H_3BO_3 (pH 9.7) as the mobile phase [586]. Chromatographic (including HSLC) and biological aspects of the phthalate esters have been reviewed by Fishbein and Albro [997].

The special advantages of TLC and LLC in the separation of lipids are discussed by Rouser [998]. The two most attractive stationary phases for HSLC of phospholipids and glycolipids were anion exchangers and silica gels. The problem of optimization of the elution parameters of lipids from HSLSC columns has been discussed by Stolyhwo and Privett [999]. The chromatographic system involved the use of a continuous series of gradient changes of pentane, ethyl ether, $CHCl_3$ and MeOH containing 8% NH_4OH with a Corasil column modified by treatment with NH_4OH. Separations were demonstrated with reference mixtures of polar and nonpolar lipids, and the technique was applied to lipids of rat blood cells. The rapid separation of nonpolar lipid classes on columns of polystyrene gel with 5% acetone in H_2O as the mobile phase was reported by Lawrence [1000]. The quantitation of the separated lipids based on cholesterol as an internal standard was carried out by means of a moving wire FID (Fig. 3.16). Glycerophosphorylcholine, phosphorylcholine, glycerophosphate, orthophosphate, and other phosphorus products from the hydrolysis of phosphatidylcholines could be separated on an anion-exchange column, and the phosphorus content of the fractions could be determined by an automatic phosphorus analysis method [1001-1004]. A two-step gradient elution of ammonium formate/sodium tetraborate and HCl was employed at a flow rate of 1 ml/min.

Katz and co-workers [1005] as well as Bhatia [929] used HSLC in water environmental studies. Chapman and Beard [1006] separated and quantified polythionates found in Wackenroder's solution. They used a column filled with activated carbon and a gradient of Na_2HPO_4/tetrahydrofuran. Doali and Juhasz [1007] applied HSLC to the qualitative analysis of compounds of propellant and explosive interest (2,4,6-trinitrotoluene and tetryl). They employed a Corasil column and a mobile phase of cyclohexane/dioxane (9:1 or 13:7). An HSLC method has been described by

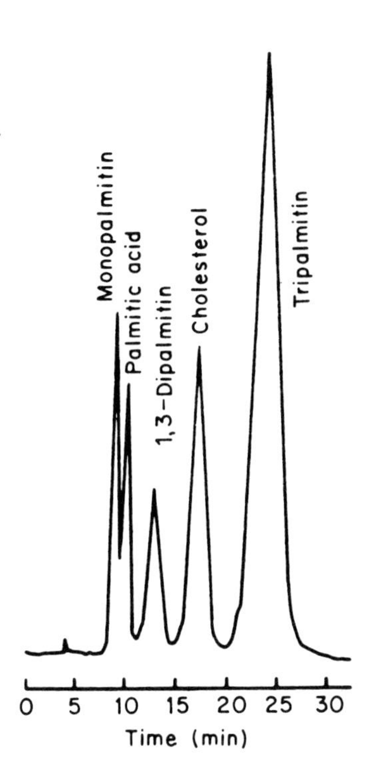

FIG. 3.16 Separation of nonpolar lipids. Column: 1 m x 4 mm, Poragel 200A; temperature: 45°C; mobile phase: 5% aqueous acetone; flow rate: 0.66 ml/min. (From Ref. 1000.)

Deelder and Hendricks [1008] for the determination of trace amounts of cyclohexanone in cyclohexanone oxime. The separation has been achieved by using a phase pair obtained from a mixture of isooctane/EtOH/H_2O. The method involved also a reaction of carbonyl-containing constituents with DNPH to yield red-colored compounds detected at 430 nm.

REFERENCES

1. Heftmann, E., Chromatography, Reinhold, New York, 1961.

2. Determann, H., in Advances in Chromatography (Giddings, J. C., and Keller, R. A., eds.), Vol. 8, Marcel Dekker, New York, 1969, p. 3.

3. Altgelt, K. H., and Segal, L., Gel Permeation Chromatography, Marcel Dekker, New York, 1971.

4. Determann, H., Gel Chromatography, Springer-Verlag, New York, 1968.

5. Giddings, J. C., Dynamics of Chromatography, Marcel Dekker, New York, 1965.

6. Martin, A. J. P., and Synge, R. L. M., Biochem. J., 35, 1358 (1941).

7. James, A. T., and Martin, A. J. P., Biochem. J., 50, 679 (1952).

8. Knox, J. H., J. Chem. Soc., 1961, 433.

9. Giddings, J. C., Anal. Chem., 35, 2215 (1963).

10. Knox, J. H., Anal. Chem., 38, 253 (1966).

11. Pretorius, V., and Smuts, T. W., Anal. Chem., 38, 274 (1966).

12. Smuts, T. W., Van Niekerk, F. A., and Pretorius, V., J. Gas Chromatogr., 5, 190 (1967).

13. Hamilton, P. B., Bogue, D. C., and Anderson, R. A., Anal. Chem., 32, 1782 (1960).

14. Karr, C., Jr., Childers, E. E., and Warner, W. C., Anal. Chem., 35, 1290 (1963).

15. Piel, E. W., Anal. Chem., 38, 60 (1966).

16. Locke, D. C., and Martire, D. E., Anal. Chem., 39, 921 (1967).

17. Locke, D. C., in Advances in Chromatography (Giddings, J. C., and Keller, R. A., eds.), Vol. 8, Marcel Dekker, New York, 1969, p. 47.

18. Hildebrand, J. H., and Scott, R. L., Solubility of Non-electrolytes, Dover, New York, 1964.

19. Huber, J. F. K., Meijers, C. A. M., and Hulsman, J. A. R. J., Anal. Chem., 44, 111 (1972).

20. Grubisic-Gallot, Z., and Benoit, H., J. Chromatogr. Sci., 9, 262 (1971).

21. Huber, J. F. K., Alderlieste, E. T., Harven, H., and Poppe, H., Anal. Chem., 45, 1337 (1973).

22. Locke, D. C., J. Chromatogr., 35, 24 (1968).

23. Knox, J. H., Ann. Rev. Phys. Chem., 24, 29 (1973).

24. Knox, J. H., Chromatogr. Newslett., 2, 1 (1973).

25. Huber, J. F. K., in Comprehensive Analytical Chemistry (Wilson, C. L., and Wilson, D. W., eds.), Vol. IIB, Elsevier, Amsterdam, 1968.

26. Done, J. N., Kennedy, G. J., and Knox, J. H., Nature, 237, 77 (1972).

27. Locke, D. C., J. Gas Chromatogr., 5, 202 (1967).

28. Horvath, C. G., and Lipsky, S. R., Anal. Chem., 39, 1893 (1967).

29. Halász, I., in Modern Practice of Liquid Chromatography (Kirkland, J. J., ed.), Wiley (Interscience), New York, 1971, p. 143.

30. Keller, R. A., J. Chromatogr. Sci., 11, 49 (1973).

31. Martire, D. E., and Locke, D. C., Anal. Chem., 43, 68 (1971).

32. Luckhurst, G. R., and Martire, D. E., Trans. Faraday Soc., 64, 1248 (1968).

33. Eon, C., Novosel, B., and Guiochon, G., J. Chromatogr., 83, 77 (1973).

34. Snyder, L. R., in Modern Practice of Liquid Chromatography (Kirkland, J. J., ed.), Wiley (Interscience), New York, 1971, p. 143.

35. Snyder, L. R., Principles of Adsorption Chromatography, Marcel Dekker, New York, 1968.

36. Kiselev, A. V., J. Chromatogr., 49, 84 (1970).

37. Kiselev, A. V., and Khopina, V. V., Trans. Faraday Soc., 65, 1936 (1969).

38. Snyder, L. R., J. Chromatogr., 36, 455 (1968).

39. Soczewinsky, E., and Golkiewicz, W., Chromatographia, 6, 269 (1973).

40. Funasaka, W., Hanai, T., Matsumoto, T., Fujimura, K., and Ando, T., J. Chromatogr., 88, 87 (1974).

41. Funasaka, W., Ando, T., Fujimura, K., and Hanai, T., Bunseki Kagaku, 20, 427 (1971).

42. Funasaka, W., Hanai, T., Fujimura, K., and Ando, T., J. Chromatogr., 72, 187 (1972).

43. Horvath, C. G., and Lipsky, S. R., J. Chromatogr. Sci., 7, 109 (1969).

44. Huber, J. F. K., J. Chromatogr. Sci., 7, 85 (1969).

45. Huber, J. F. K., and Hulsman, J. A. R. J., Anal. Chim. Acta, 38, 305 (1967).

46. Deelder, R. S., Hendricks, P. J. H., and Kroll, M. G. F., J. Chromatogr., 57, 67 (1971).

47. Karger, B. L., in Modern Practice of Liquid Chromatography (Kirkland, J. J., ed.), Wiley (Interscience), New York, 1971, p. 29.

48. Horne, D. S., Knox, J. H., and McLaren, L., Separation Sci., 1, 531 (1966).

49. Martin, M., Blu, G., and Guiochon, G., J. Chromatogr. Sci., 11, 641 (1973).

50. Loheac, J., Martin, M., and Guiochon, G., Analusis, 1, 9 (1972).

51. Deninger, G., Ber. Bunseges Phys. Chem., 77, 145 (1973).

52. Hamilton, P. B., in Advances in Chromatography (Giddings, J. C., and Keller, R. A., eds.), Vol. 2, Marcel Dekker, New York, 1966.

53. Halász, I., and Walking, P., J. Chromatogr. Sci., 7, 129 (1969).

54. Knox, J. H., and Parcher, J. F., Anal. Chem., 41, 1599 (1969).

55. Snyder, L. R., J. Chromatogr. Sci., 7, 352 (1969).

56. Snyder, L. R., Anal. Chem., 39, 698 (1967).

57. Waters, J. L., Little, J. N., and Horgan, D. F., J. Chromatogr. Sci., 7, 293 (1969).

58. Simpson, D. W., and Wheaton, R. M., Chem. Eng. Progr., 50, 45 (1954).

59. Kirkland, J. J., J. Chromatogr. Sci., 10, 129 (1972).

60. Kirkland, J. J., J. Chromatogr. Sci., 10, 593 (1972).

61. Jardy, A., and Rosset, R., J. Chromatogr., 83, 195 (1973).

62. Morris, B. M., Mode, V. A., and Sisson, D. H., J. Chromatogr., 71, 389 (1972).

63. Beachell, H. C., and De Stefano, J. J., J. Chromatogr. Sci., 10, 481 (1972).

64. De Stefano, J. J., and Beachell, H. C., J. Chromatogr. Sci., 10, 654 (1972).

65. Knox, J. H., and Salem, M., J. Chromatogr. Sci., 7, 745, (1969).

66. Snyder, L. R., J. Chromatogr. Sci., 10, 200 (1972).

67. Snyder, L. R., J. Chromatogr. Sci., 10, 369 (1972).

68. Smuts, T. W., D.Sc. thesis, Univ. of Pretoria, 1967.

69. Smuts, T. W., and Pretorius, V., Anal. Chem., 44, 121 (1972).

70. Sie, S. T., and Rijnders, G. W. A., Anal. Chim. Acta, 38, 3 (1967).

71. Giddings, J. C., J. Chem. Ed., 35, 588 (1958).

72. Giddings, J. C., Nature, 184, 357 (1959).

73. Scott, R. P. W., Blackburn, D. W. J., and Wilkins, T., in Advances in Gas Chromatography 1967 (Zlatkis, A., ed.), Preston, Evanston, Ill., 1967, p. 160.

74. Walkling, P., Ph.D. thesis, Univ. Frankfurt/Main, 1968.

75. Scott, R. P. W., and Lawrence, J. G., in Advances in Chromatography 1969 (Zlatkis, A., ed.), Preston, Evanston, Ill., 1969, p. 276.

76. Halász, I., and Walkling, P., in Advances in Chromatography 1969 (Zlatkis, A., ed.), Preston, Evanston, Ill., 1969, p. 310.

77. Naefe, M., Ph.D. thesis, Univ. Frankfurt/Main, 1970.

78. Horne, D. S., Knox, J. H., and McLaren, L., in Separation Techniques in Chemistry and Biochemistry (Keller, R. A., ed.), Marcel Dekker, New York, 1967, p. 97.

79. Stewart, H. N. M., Amos, R., and Perry, S. G., J. Chromatogr., 38, 209 (1968).

80. Halász, I., and Naefe, M., Anal. Chem., 44, 76 (1972).

81. Jardy, A., and Rosset, R., Bull. Soc. Chim. Fr., 1973, 156.

82. Snyder, L. R., in Gas Chromatography 1970 (Stock, R., and Perry, S., eds.), Inst. of Petroleum, London, 1971, p. 81.

83. Halász, I., Kroneisen, A., Gerlach, H. O., and Walkling, P., Z. Anal. Chem., 234, 81 (1968).

84. Majors, R. E., J. Chromatogr. Sci., 11, 88 (1973).

85. Done, J. N., and Knox, J. H., J. Chromatogr. Sci., 10, 606 (1972).

86. Klein, P., Anal. Chem., 33, 1737 (1961).

87. Kiselev, A. V., Frolov, I., and Yashin, Y. I., in 5th Intern. Symp. on Column Liquid Chromatography (Kovats, E., ed.), Lausanne, 1969, p. 116.

88. Kirkland, J. J., J. Chromatogr. Sci., 7, 7 (1969).

89. Geiss, F., Schlitt, H., and Klose, A., Z. Anal. Chem., 213, 321 (1965).

90. Engelhardt, H., and Wiedemann, H., Anal. Chem., 45, 1641 (1973).

91. Halász, I., and Naefe, M., in Advances in Chromatography 1971 (Zlatkis, A., ed.), Univ. of Houston, Texas, 1971, p. 195.

92. Rajcsanyi, P. M., Unpublished results.

93. Sie, S. T., and Van den Hoed, N., J. Chromatogr. Sci., 7, 257 (1969).

94. De Stefano, J. J., and Beachell, H. C., J. Chromatogr. Sci., 8, 434 (1970).

95. Frolov, I. I., Vorobyeva, R. G., Mironova, I. V., Chernov, A. Z., and Yashin, Y. I., J. Chromatogr., 80, 167 (1973).

96. Randau, D., and Schnell, W., J. Chromatogr., 57, 373 (1971).

97. Engelhardt, H., and Weigand, N., Anal. Chem., 45, 1149 (1973).

98. Halász, I., Engelhardt, H., Asshauer, J., and Karger, B. L., Anal. Chem., 42, 1460 (1970).

99. Snyder, L. R., Advan. Anal. Chem. Instrum., 3, 251 (1964).

100. Snyder, L. R., J. Chromatogr. Sci., 9, 322 (1971).

101. Deininger, G., and Halász, I., J. Chromatogr. Sci., 9, 83 (1971).

102. Halász, I., Gerlach, H. O., Kroneisen, A., and Walkling, P., Z. Anal. Chem., 234, 97 (1968).

103. Versino, B., and Schlitt, H., Chromatographia, 5, 332, (1972).

104. Schlitt, H., and Geiss, F., J. Chromatogr., 67, 261 (1972).

105. Hawkes, S. J., J. Chromatogr. Sci., 7, 526 (1969).

106. Smuts, T. W., De Clerk, K., and Pretorius, V., Separation Sci., 3, 43 (1968).

107. Giddings, J. C., Anal. Chem., 36, 741 (1964).

108. Giddings, J. C., Anal. Chem., 35, 439 (1963).

109. Hawkes, S. J., J. Chromatogr., 68, 1 (1972).

110. Knox, J. H., and Saleem, M., J. Chromatogr. Sci., 7, 614 (1969).

111. Majors, R. E., and MacDonald, F. R., J. Chromatogr., 83, 169 (1973).

112. Karger, B. L., in Modern Practice of Liquid Chromatography (Kirkland, J. J., ed.), Wiley (Interscience), New York, 1971, p. 3.

113. Snyder, L. R., Anal. Chem., 39, 705 (1967).

114. Grushka, E., J. Chromatogr. Sci., 10, 616 (1972).

115. Jolley, R. L., and Scott, C. D., J. Chromatogr., 47, 272 (1970).

116. Chilcote, D. D., and Scott, C. D., Anal. Chem., 45, 721 (1973).

117. Chilcote, D. D., Clin. Chem., 19, 826 (1973).

118. Scott, C. D., Jansen, J. M., and Pitt, W. W., Jr., Amer. J. Clin. Pathol., 53, 739 (1970).

119. Butts, W. C., Anal. Biochem., 46, 187 (1972).

120. Scott, R. P. W., J. Chromatogr. Sci., 9, 449 (1971).

121. Scott, R. P. W., J. Chromatogr. Sci., 10, 189 (1972).

122. Snyder, L. R., J. Chromatogr. Sci., 10, 187 (1972).

123. Golay, M. J. E., Chromatographia, 6, 242 (1973).

124. Halász, I., Chromatographia, 6, 481 (1973).

125. Rajcsanyi, P. M., and Ötvös, L., Separation Purification Meth., 2, 361 (1973).

126. Arikawa, Y., and Toshida, K., Hitachi Rev., 16, 236 (1967).

127. Bombaugh, K. J., J. Chromatogr., 53, 27 (1970).

128. Bombaugh, K. J., Amer. Lab., 1969, July.

129. Bombaugh, K. J., Can. Res. Devel., 1969, Sept./Oct.

130. Cooper, A. R., Chem. Britain, 9, 112 (1973).

131. Ecker, E., Chem. Ztg., 11, 511 (1971).

132. Fallick, G., Amer. Lab., 1973, May.

133. Guiochon, G., Ber. Bunseges Phys. Chem., 77, 207 (1973).

134. Hatano, H., and Watanabe, S., Kagaku No Ryoiki, 25, 824 (1971).

135. Hatano, H., Res. Devel., 24, 28 (1973).

136. Hirata, Y., Farumashia, 8, 756 (1972).

137. Huber, J. F. K., Ber. Bunseges Phys. Chem., 77, 179 (1973).

138. Kahn, H. L., and Bitterfield, Z., Ind. Res., 15, 32 (1973).

139. Karasek, F. W., Res. Devel., 24, 52 (1973).

140. Kawamura, J., Japan Analyst, 22, 229 (1973).

141. Otocka, E. P., Acc. Chem. Res., 6, 348 (1973).

142. Simpson, D., Lab. Pract., 22, 593 (1973).

143. Schill, G., Kem. Tidskr., 84, 42 (1972).

144. Vigh, Gy., and Inczedy, J., Magy. Kem. Lapja, 28, 113 (1973).

145. Vigh, Gy., and Inczedy, J., Magy. Kem Lapja, 28, 577 (1973).

146. Vanheertum, R., Meded. Vlaam. Chem. Ver., 33, 173 (1971).

147. Yamabe, T., Yuki Gosei Kagaku Kyoai Shi, 31, 115 (1973).

148. Ali, S. L., Pharm. Ztg., 118, 1139 (1973).

149. Done, N., and Knox, J. H., Process. Biochem., 7, 11 (1972).

150. Eisenbeiss, F., Chem. Ztg., 95, 237 (1971).

151. Fallick, G., and Zenie, F. H., Can. Res. Develop., 6, 24 (1973).

152. Leicht, R. E., and Kirkland, J. J., Ind. Res., 1970, 36.

153. Yost, R. W., and Conlon, R. D., Chromatogr. Newslett., 1, 41 (1972).

154. Bombaugh, K. J., in Progress in Analytical Chemistry (Simmons, I. L., and Ewing, G. W., eds.), Vol. 6, Plenum, New York, 1973, p. 193.

155. Culberson, C. F., Bryologist, 75, 54 (1972).

156. Bombaugh, K. J., Amer. Lab., 1973, May.

157. Chandler, D. C., and McNair, H. M., J. Chromatogr. Sci., 11, 468 (1973).

158. Gilding, D. K., Amer. Lab., 1969, Oct.

159. Knox, J. H., Lab. Pract., 22, 52 (1973).

160. Leslie, R. C., Food Manuf., 47, 41 (1972).

161. Price, L. W., Lab. Pract., 21, 661 (1972).

162. Veening, H., J. Chem. Ed., 43, A429 (1973).

163. Veening, H., J. Chem. Ed., 43, A529 (1973).

164. Martin, M., and Guiochon, G., Bull. Soc. Chim. Fr., 1973, 168.

165. Dietsch, G., and Ecker, E., Siemens Ztg., 47, 483 (1973).

166. Kochen, W., and Trefz, F., Z. Anal. Chem., 261, 342 (1972).

167. Ecker, E., Messtechnik (Brunswick), 80, 307 (1972).

168. Hatano, H., Hitachi Sci. Instrum. News, 5, 14 (1972).

169. Schrenker, H., Cz. Chemie-Techn., 1, 73 (1972).

170. Stahl, K. W., Schuppe, E., and Potthast, H., GIT Fachz. Lab., 17, 563 (1973).

171. Felton, H., J. Chromatogr. Sci., 7, 13 (1969).

172. Henry, R. A., in Modern Practice of Liquid Chromatography (Kirkland, J. J., ed.), Wiley (Interscience), New York, 1971, p. 55.

173. Hupe, K. P., and Schrenker, H., Chromatographia, 5, 44 (1972).

174. Allington, W. B., ISCO Appl. Res. Bull., No. 6, 1971.

175. Snyder, L. R., Chromatogr. Rev., 7, 1 (1965).

176. Snyder, L. R., J. Chromatogr. Sci., 8, 692 (1970).

177. Snyder, L. R., and Saunders, D. L., J. Chromatogr. Sci., 7, 195 (1969).

178. Scott, R. P. W., and Lawrence, J. G., J. Chromatogr. Sci., 8, 619 (1970).

179. Scott, R. P. W., and Lawrence, J. G., J. Chromatogr. Sci., 7, 65 (1969).

180. Maggs, R. J., and Young, T. E., in Gas Chromatography 1968 (Harbourne, C. L. A., ed.), Inst. of Petroleum, London, 1969, p. 217.

181. Maggs, R. J., J. Chromatogr. Sci., 7, 145 (1969).

182. Scott, R. P. W., J. Chromatogr. Sci., 9, 385 (1971).

183. Scott, R. P. W., and Kucera, P., Anal. Chem., 45, 749 (1973).

184. Scott, R. P. W., and Kucera, P., J. Chromatogr. Sci., 11, 83 (1973).

185. Scott, R. P. W., and Kucera, P., J. Chromatogr., 83, 257 (1973).

186. Anon., Chem. Eng. News, 1973, July, 9.

187. Byrne, S. H., Schmit, J. A., and Johnson, P. E., J. Chromatogr. Sci., 9, 592 (1971).

188. Delfel, N. E., Anal. Chem., 38, 1429 (1966).

189. Chilcote, D. D., Scott, C. D., and Pitt, W. W., Jr., J. Chromatogr., 75, 175 (1973).

190. Scott, C. D., Jolley, R. L., Pitt, W. W., Jr., and Johnson, W. F., Amer. J. Clin. Pathol., 53, 701 (1970).

191. Solms, D. J., Smuts, T. W., and Pretorius, V., J. Chromatogr. Sci., 9, 600 (1971).

192. Bombaugh, K. J., King, R. N., and Cohen, A. J., J. Chromatogr., 43, 332 (1969).

193. Nester/Faust -Perkin Elmer Appl. Bull., 1969/1.

194. Conlon, R. D., Pittsburgh Conference on Analytical Chemistry, Cleveland, Feb. 28-March 5, 1971.

195. Katz, S., J. Chromatogr., 53, 415 (1970).

196. Snyder, L. R., J. Chromatogr. Sci., 7, 595 (1969).

197. Jentoft, R. E., and Gouw, T. H., Anal. Chem., 38, 949 (1966).

198. Bombaugh, K. J., Levangie, R. F., King, R. N., and Abrahams, L., J. Chromatogr. Sci., 8, 657 (1970).

199. Micrometrics Instrument Co., Bull., 1973.

200. Du Pont LC Product Bull., 840PB1, 1971.

201. Waters Associates, Bull., 72-207, 1972.

202. Bidlingmeyer, B. A., Hooker, R. P., Lochmüller, C. H., and Rogers, L. B., Separation Sci., 4, 439 (1969).

203. Bidlingmeyer, B. A., and Rogers, L. B., Anal. Chem., 43, 1882 (1971).

204. Bonneleycke, B. E., J. Chromatogr., 45, 135 (1969).

205. Machin, A. F., Morris, C. R., and Quick, M. P., J. Chromatogr., 72, 388 (1972).

206. Sie, S. T., and Rijnders, W. A., Separation Sci., 2, 729 (1967).

207. Jentoft, R. E., and Gouw, T. H., J. Chromatogr. Sci., 8, 138 (1970).

208. Deininger, G., and Halász, I., J. Chromatogr., 60, 65 (1971).

209. Halász, I., and Deininger, G., Anal. Chem., 228, 321 (1967).

210. Karger, B. L., and Berry, L. V., Anal. Chem., 44, 93 (1972).

211. MacDonald, F. R., Intern. Lab., 1973, July/August.

212. Waters Associates, Bull., PB 73-210, 1973.

213. Pound, N. J., Sears, R. W., and Butterfield, A. G., Anal. Chem., 45, 1001 (1973).

214. Rössler, G., Schneider, W., and Halász, I., Chromatographia, 6, 237 (1973).

215. Berry, L. V., and Karger, B. L., Anal. Chem., 45, 819A (1973).

216. MacDonald, F. R., and Munk, M. N., California ACS Summer Round Table, 1970, July.

217. Rajcsanyi, P. M., Unpublished results.

218. Beckman Instruments, Brit. Pat., 1,263,481.

219. Divelbiss, H. N., and Goostree, B. E., Ger. Pat., 2,040,835.

220. du Pont LC Product Bull., 830PB4, 1971.

221. Scott, C. D., Johnson, W. T., and Walker, V. E., Anal. Biochem., 32, 182 (1969).

222. Jentoft, R. E., and Gouw, T. H., Anal. Chem., 40, 923 (1968).

223. Siemens Bull., E 634/1006-101, 1973.

224. Fleischer, J., Chromatographia, 7, 80 (1974).

225. Smuts, T. W., Solms, D. J., Van Niekerk, F. A., and Pretorius, V., J. Chromatogr. Sci., 7, 24 (1969).

226, Krejci, M., Vespalec, R., and Sirec, M., J. Chromatogr., 65, 333 (1972).

227. Cassidy, R. M., and Frei, R. W., Anal. Chem., 44, 2250 (1972).

228. Lindley, H., Cranston, R. W., Sutherland, W. J. A., and Ritchie, C. L., J. Chromatogr., 86, 178 (1973).

229. Karasek, F. W., Res. Devel., 24, 54 (1973).

230. Kirkland, J. J., J. Chromatogr. Sci., 7, 361 (1969).

231. Karger, B. L., and Barth, H., Anal. Lett., 4, 595 (1971).

232. Scott, R. P. W., Blackburn, D. W., and Wilkins, T., J. Gas Chromatogr., 5, 183 (1967).

233. Barth, H., Dallmeier, E., and Karger, B. L., Anal. Chem., 44, 1726 (1972).

234. Whitlock, L. R., Porter, R. S., and Johnson, J. F., J. Chromatogr. Sci., 9, 437 (1971).

235. Koutsky, J. A., and Adler, R. J., Can. J. Chem. Eng., 43, 239 (1964).

236. Halász, I., and Walkling, P., Ber. Bunseges Phys. Chem., 74, 66 (1970).

237. Heitz, W., J. Chromatogr., 83, 223 (1973).

238. Baumann, F., and Hadden, N., in Basic Liquid Chromatography, Varian, Walnut Creek, 1972, pp. 2-6.

239. Varian, Series 4100/4200 LC, Bull., 1972.

240. Scott, C. D., in Modern Practice of Liquid Chromatography (Kirkland, J. J., ed.), Wiley (Interscience), New York, 1971, p. 292.

241. Scott, R. P. W., and Kucera, P., J. Chromatogr. Sci., 9, 641 (1971).

242. Wohlleben, G., Cz. Chem.-Techn., 1, 81 (1972).

243. Bombaugh, K. J., Dark, W. A., and Little, J. N., Anal. Chem., 41, 1337 (1969).

244. Moore, J. C., U.S. Pat. 3,326,875.

245. Bombaugh, K. J., in Modern Practice of Liquid Chromatography (Kirkland, J. J., ed.), Wiley (Interscience), New York, 1971, p. 249.

246. Little, J. N., Waters, J. L., Bombaugh, K. J., and Pauplis, W. J., in 5th Intern. Symp. on Column Liquid Chromatography (Kovats, E., ed.), Lausanne, 1969, p. 128.

247. Cantow, M. J. R., and Johnson, J. F., J. Appl. Polym. Sci., 11, 1851 (1967).

248. Ross, J. H., and Casto, M. E., J. Polym. Sci. Part C, 21, 143 (1968).

249. Chang, T. L., Anal. Chem., 40, 989 (1968).

250. Haller, W., Nature, 206, 693 (1965).

251. Telepchak, M. J., J. Chromatogr., 83, 125 (1973).

252. Cooper, A. R., Bruzzone, A. R., Cain, J. H., and Barrall, E. M., J. Appl. Polym. Sci., 15, 571 (1971).

253. Cooper, A. R., and Johnson, J. F., J. Appl. Polym. Sci., 13, 1487 (1969).

254. Kirkland, J. J., J. Chromatogr. Sci., 9, 206 (1971).

255. Huber, J. F. K., Chimia, Suppl., 1970, 24.

256. Scott, C. D., and Lee, N. E., J. Chromatogr., 42, 263 (1969).

257. Scott, C. D., in Modern Practice of Liquid Chromatography (Kirkland, J. J., ed.), Wiley (Interscience), New York, 1971, p. 309.

258. Majors, R. E., Anal. Chem., 44, 1722 (1972).

259. Stubert, W., Chromatographia, 6, 50 (1973).

260. Kirkland, J. J., Anal. Chem., 43, 36A (1971).

261. Majors, R. E., Amer. Lab., 1972, May.

262. Leicht, R. E., and De Stefano, J. J., J. Chromatogr. Sci., 11, 105 (1973).

263. Electro Nucleonics, CPG Bull., 1972.

264. Horvath, C. G., Preiss, B., and Lipsky, S. R., Anal. Chem., 39, 1422 (1967).

265. Du Pont LC Product Bull., 820M9, 1970.

266. Du Pont LC Product Bull., 820PB4, 1971.

267. Durrum Resin Report No. 2, 1971.

268. Gas-Chrom News Letter, 1972/5.

269. Horvath, C. G., Ion Exch. Solvent Extr., 5, 207 (1973).

270. Bly, D. L., Science, 168, 527 (1970).

271. Pidacks, C., J. Chromatogr. Sci., 8, 618 (1970).

272. Dark, W. A., and Limpert, R. J., J. Chromatogr. Sci., 11, 114 (1973).

273. Kennedy, G. J., and Knox, J. H., J. Chromatogr. Sci., 10, 549 (1972).

274. Little, J. N., Horgan, D. F., and Bombaugh, K. J., J. Chromatogr. Sci., 8, 625 (1970).

275. De Vries, A. J., Le Page, M., Beau, R., and Guillemin, C. L., Anal. Chem., 39, 935 (1967).

276. Unger, K., Ringe, P., Schick-Kalb, J., and Straube, B., Z. Anal. Chem., 264, 267 (1973).

277. Kirkland, J. J., J. Chromatogr., 83, 149 (1973).

278. Knox, J. H., and Vasvari, G., J. Chromatogr., 83, 181 (1973).

279. Scott, C. D., Anal. Biochem., 24, 292 (1968).

280. Unger, K., Schick-Kalb, J., and Krebs, K. F., J. Chromatogr., 83, 5 (1973).

281. Tesarik, K., and Necasova, M., J. Chromatogr., 75, 1 (1973).

282. Tesarik, K., and Pracharova, M., J. Chromatogr., 84, 225 (1973).

283. Otocka, E. P., Anal. Chem., 45, 1969 (1973).

284. Abel, E. W., Pollard, F. H., Uden, P. C., and Nickless, G., J. Chromatogr., 22, 23 (1966).

285. Karasek, F. W., Res. Devel., 22, 34 (1971).

286. Kirkland, J. J., U.S. Pat. 3,488,922.

287. Gilpin, R. K., and Burke, M. F., Anal. Chem., 45, 1383 (1973).

288. Stewart, H. N. M., and Perry, S. G., J. Chromatogr., 37, 97 (1968).

289. Bossart, C. J., U.S. Pat. 3,514,925.

290. Stehl, R. H., U.S. Pat. 3,664,967.

291. Halász, I., and Sebestian, I., Angew. Chem., Int. Ed., 8, 453 (1969).

292. Aue, W., and Hastings, C. R., J. Chromatogr., 42, 319 (1969).

293. Hastings, C. R., Aue, W., and Angl, J. M., J. Chromatogr., 53, 487 (1970).

294. Kirkland, J. J., and De Stefano, J. J., J. Chromatogr. Sci., 8, 309 (1970).

295. Locke, D. C., Schmermund, J. T., and Banner, B., Anal. Chem., 44, 90 (1972).

296. Chromatography Notes, 1972, Dec.

297. Chromatography Notes, 1973, June.

298. Chromatography Notes, 1973, Sept.

299. Waters Associates, Bull. DS 72-015, 1972.

300. Brust, O. E., Sebestian, I., and Halász, I., J. Chromatogr., 83, 15 (1973).

301. Unger, K., Thomas, W., and Adrian, P., Kolloid-Z. Polym., 251, 45 (1973).

302. Unger, K., Berg, K., and Gallei, E., Kolloid-Z. Polym., 234, 1108 (1969).

303. Unger, K., Berg, K., Gallei, E., and Erdel, G., Fortschr. Kolloid. Polym., 55, 34 (1971).

304. Berg, K., and Unger, K., Kolloid-Z. Polym., 246, 682 (1971).

305. Unger, K., Schier, G., and Beisel, V., Chromatographia, 6, 456 (1973).

306. Hill, J. M., J. Chromatogr., 76, 455 (1973).

307. Novotny, M., Bektesh, S. L., Denson, K. B., Grohmann, K., and Parr, W., Anal. Chem., 45, 971 (1973).

308. Novotny, M., Bektesh, S. L., and Grohmann, K., J. Chromatogr., 83, 25 (1973).

309. Grushka, E., and Scott, R. P. W., Anal. Chem., 45, 1626 (1973).

310. Ray, S., and Frei, R. W., J. Chromatogr., 71, 451 (1972).

311. Rabel, F. M., Anal. Chem., 45, 957 (1973).

312. McNair, H. M., and Chandler, C. D., Anal. Chem., 45, 1117 (1973).

313. Chandler, C. D., Diss. Abstr. Int. B, 34, 1407 (1973).

314. Ross, W. D., and Jefferson, R. T., J. Chromatogr. Sci., 8, 386 (1970).

315. Karger, B. L., Conroe, K., and Engelhardt, H., J. Chromatogr. Sci., 8, 242 (1970).

316. Locke, D. C., J. Chromatogr. Sci., 11, 120 (1973).

317. du Pont LC Product Bull., 820PB5, 1971.

318. Horvath, C. G., in The Practice of Gas Chromatography (Ettre, L. S., and Zlatkis, A., eds.), Interscience, New York, 1967, p. 201.

319. Majors, R. E., Anal. Chem., 45, 755 (1973).

320. Karger, B. L., Engelhardt, H., Conroe, K., and Halász, I., in Gas Chromatography 1970 (Stock, R., and Perry, S. G., eds.), Inst. of Petroleum, London, 1971, p. 112.

321. Kirkland, J. J., and Dilks, C. H., Jr., Anal. Chem., 45, 1778 (1973).

322. Huber, J. F. K., J. Chromatogr. Sci., 9, 72 (1971).

323. Horgan, D. F., and Little, J. N., J. Chromatogr. Sci., 10, 76 (1972).

324. Baumann, F., and Hadden, N., in Basic Liquid Chromatography, Varian, Walnut Creek, 1972, p. 2-2.

325. Leicht, R., J. Chromatogr. Sci., 9, 531 (1971).

326. Shmukler, H. W., J. Chromatogr. Sci., 8, 581 (1970).

327. Snyder, L. R., in Modern Practice of Liquid Chromatography (Kirkland, J. J., ed.), Wiley (Interscience), New York, 1971, p. 215.

328. Snyder, L. R., Principles of Adsorption Chromatography, Marcel Dekker, New York, 1968, p. 78.

329. Scott, C. D., in Modern Practice of Liquid Chromatography (Kirkland, J. J., ed.), Wiley (Interscience), New York, 1971, p. 287.

330. Inczedy, J., Analytical Applications of Ion Exchangers, Pergamon, London, 1966.

331. Strelow, F. W. E., Liebnberg, C. J., and Von S. Toerin, F., Anal. Chim. Acta, 43, 465 (1968).

332. Quano, A., and Biesenberg, J. A., J. Chromatogr. Sci., 9, 193 (1971).

333. Determann, H., and Lampert, K., J. Chromatogr., 69, 123 (1972).

334. Van Kreveld, M. E., and Van den Hoed, N., J. Chromatogr., 83, 111 (1973).

335. Hesse, G., and Engelhardt, H., J. Chromatogr., 21, 228 (1966).

336. Giddings, J. C., Myers, N. N., McLaren, L., and Keller, R. A., Science, 162, 67 (1968).

337. Keller, R. A., and Snyder, L. R., J. Chromatogr. Sci., 9, 346 (1971).

338. Kirkland, J. J., in Modern Practice of Liquid Chromatography (Kirkland, J. J., ed.), Wiley (Interscience), New York, 1971, p. 181.

339. Guiochon, G., Chromatographia, 5, 571 (1972).

340. Krejci, M., and Pospisilova, N., J. Chromatogr., 73, 105 (1972).

341. Veening, H., J. Chem. Ed., 47, A549, A675, A749 (1970).

342. Veening, H., J. Chem. Ed., 50, A481 (1973).

343. Huber, J. F. K., J. Chromatogr. Sci., 7, 172 (1969).

344. Conlon, R. D., Anal. Chem., 41, 107A (1969).

345. Munk, M. N., J. Chromatogr. Sci., 8, 491 (1970).

346. Oster, H., Messtechnik (Brunswick), 80, 375 (1972).

347. Martin, A. J. P., Pure Appl. Chem., 34, 83 (1973).

348. Polesuk, J., and Howery, D. G., J. Chromatogr. Sci., 11, 226 (1973).

349. Deininger, G., Kroneisen, A., and Halász, I., Chromatographia, 3, 329 (1970).

350. Conlon, R. D., in Instrumentation in Analytical Chemistry (Senzel, A. J., ed.), Amer. Chem. Soc., Washington, 1973, p. 188.

351. Hupe, K. P., Angew. Chem., 82, 12 (1970).

352. Byrne, S. H., in Modern Practice of Liquid Chromatography (Kirkland, J. J., ed.), Wiley (Interscience), New York, 1971, p. 97.

353. Hartmann, C. H., Anal. Chem., 43, 113A (1971).

354. Siemens, Spektralphotometer PM4, Bull., 1972.

355. Carr, D., Varian Instr. Appl., 7, 14 (1973).

356. du Pont LC Product Bull., 820PB6-E, 1972.

357. du Pont LC Product Bull., 410PB1, 1972.

358. du Pont, 410 Precision Photometer, Bull., 1972.

359. ISCO, Catalog, UA-5 UV Monitor, 1973.

360. Cecil Instruments, CE 212 Bull., 1973.

361. Chromatec, Model 800 UV detector, Bull., 1973.

362. Gilson, Recording UV spectrophotometer, Bull., 1973.

363. Schoeffel, Model 770 UV-VIS detector, Bull., 1973.

364. Munk, M. N., Pittsburgh Conference on Analytical Chemistry, Cleveland, March, 1973.

365. Laboratory Data Control, 1205 UV detector, Bull., 1972.

366. Laboratory Data Control, 1285 UV detector, Bull., 1973.

367. Perkin-Elmer-Nester/Faust, Bull. LC-8, 1972.

368. Chromatronix, Inc., Bull. NPB 472, 1972.

369. Chromatronix, Inc., Bull. CHH 772, 1972.

370. Waters Associates, Inc., Bull. AN 71-107, 1971.

371. Waters Associates, Inc., Bull. DS 72-012, 1972.

372. KNAUER, UV-photometer, Bull., 1973.

373. Molecular Separations, Inc., Bull. No. 501, 1973.

374. Chromatec, Inc., Bull. DS 102, 1973.

375. Chromatec, Inc., Bull. LC-2200, 1973.

376. Watson, E. S., Ger. Pat. 2,258,208.

377. Spackmann, D. H., Stein, W. H., and Moore, S., Anal. Chem., 30, 1190 (1958).

378. Horvath, C. G., and Lipsky, S. R., Nature, 211, 748 (1966).

379. Brdicka, R., Chem Listy., 65, 872 (1971).

380. Esser, R. J. E., Z. Anal. Chem., 236, 59 (1968).

381. Kelemen, S. P., and Degens, E. T., Nature, 211, 857 (1966).

382. Kirkland, J. J., Anal. Chem., 40, 391 (1968).

383. Hamilton, P. B., Rev. Sci. Instrum., 38, 1301 (1967).

384. Hamilton, P. B., Ann. N.Y. Acad. Sci., 102, 55 (1962).

385. Hamilton, P. B., Anal. Chem., 35, 2055 (1963).

386. Hamilton, P. B., Fed. Proc., 24, 656 (1965).

387. Thacker, L. H., Scott, C. D., and Pitt, W. W., Jr., J. Chromatogr., 51, 175 (1970).

388. Thacker, L. H., Pitt, W. W., Jr., Katz, S., and Scott, C. D., Clin. Chem., 16, 626 (1970).

389. Brooker, G., Anal. Chem., 43, 1095 (1971).

390. Callmer, K., and Nilsson, O., Chromatographia, 6, 517 (1973).

391. Bylina, A., Sybilska, D., Grabowski, Z. R., and Koszewski, J., J. Chromatogr., 83, 357 (1973).

392. Maley, L. E., J. Chem. Ed., 45, A467 (1968).

393. Watson, E. S., Amer. Lab., 1969, Sept.

394. Siemens, Diff. Refractometer, Bull., 1972.

395. du Pont LC Product Bull., 830PB2, 1971.

396. du Pont LC Product Bull., 820PB1, 1970.

397. KNAUER, Diff. Refractometer, Bull., 1973.

398. Waters Associates, Bull., PB 72-208, 1972.

399. Waters Associates, Bull., AN 71-101, 1971.

400. Johnson, H. W., Jr., Campanile, V. A., and LeFebre, H. A., Anal. Chem., 39, 32 (1967).

401. Deininger, G., and Halász, I., J. Chromatogr. Sci., 8, 499 (1970).

402. Deininger, G., and Halász, I., in Advances in Chromatography 1970 (Zlatkis, A., ed.), Univ. of Houston, Texas, 1971, p. 336.

403. Bakken, M., and Stenberg, V. I., J. Chromatogr. Sci., 9, 603 (1971).

404. KNAUER, UV-RI dual detector, Bull., 1973.

405. Gow Mac Instrument Co., Bull., SB-80, 1973.

406. Haahti, E., and Nikarri, T., Acta Chem. Scand., 17, 2565 (1963).

407. James, A. T., Ravenhill, J. R., and Scott, R. P. W., Chem. Ind., 18, 746 (1964).

408. Karmen, A., Anal. Chem., 38, 286 (1966).

409. Fowlis, I. A., Maggs, R. J., and Scott, R. P. W., J. Chromatogr., 15, 471 (1964).

410. Young, T. E., and Maggs, R. J., Anal. Chim. Acta, 38, 105 (1967).

411. Maggs, R. J., Chromatographia, 1, 43 (1968).

412. JEOL, Bull., LC-OR 105, 1972.

413. Nota, G., and Palombari, R., J. Chromatogr., 62, 153 (1971).

414. Dubsky, H., Pajurek, J., and Krejci, M., Chem. Listy, 66, 93 (1972).

415. Burnev, N. P., USSR Pat. 368,542.

416. Karmen, A., in Separation Techniques in Chemistry and Biochemistry (Keller, R. A., ed.), Marcel Dekker, New York, 1967, p. 345.

417. Pattison, M. H., Intern. Lab., 4, 50 (1972).

418. Stevens, R. H., J. Gas Chromatogr., 6, 375 (1968).

419. Scott, R. P. W., and Lawrence, J. G., J. Chromatogr. Sci., 8, 65 (1970).

420. Scott, R. P. W., Brit. Pat. 1,045,801.

421. Scott, R. P. W., Brit. Pat. 998,107.

422. Cropper, F. R., and Heinekey, D. M., Column, 1967/2.

423. Maggs, R. J., Column, 1967/2.

424. Pye, M. J., Column, 1972/14.

425. Van Dijk, J. H., J. Chromatogr. Sci., 10, 31 (1972).

426. Lapidus, B. M., and Karmen, A., J. Chromatogr. Sci., 10, 103 (1972).

427. Stouffer, J. E., Kersten, T. E., and Kruger, P. M., Biochim. Biophys. Acta, 93, 191 (1964).

428. Stouffer, J. E., Oakes, P. L., and Schlatter, J. E., J. Gas Chromatogr., 4, 89 (1966).

429. Pretorius, V., and Van Rensburg, J. F. J., J. Chromatogr. Sci., 11, 355 (1973).

430. Stolyhwo, A., Privett, O. S., and Erdahl, W. L., J. Chromatogr. Sci., 11, 263 (1973).

431. Johnson, H. W., Jr., Seibert, E. E., and Stross, F. H., Anal. Chem., 40, 403 (1968).

432. Coll, H., Johnson, H. W., Jr., Polgar, A. G., Seibert, E. E., and Stross, F. H., J. Chromatogr. Sci., 7, 30 (1969).

433. Dubsky, H., J. Chromatogr., 71, 395 (1972).

434. Dubsky, H., Chem. Listy., 67, 533 (1973).

435. Scott, R. P. W., Anal. Chem., 35, 481 (1963).

436. Stafford, D. T., Ph.D. thesis, Virginia Polytechnical Institute and State University, 1972.

437. Scott, R. P. W., J. Chromatogr. Sci., 11, 349 (1973).

438. Smuts, T. W., Richter, P. W., and Pretorius, V., J. Chromatogr. Sci., 9, 457 (1971).

439. Claxton, G. C., J. Chromatogr., 2, 136 (1959).

440. Glumer, M., Anal. Chem., 32, 772 (1960).

441. Naono, T., Pittsburgh Conference on Analytical Chemistry, Cleveland, 1964, March.

442. Cashaw, J. L., Segura, R., and Zlatkis, A., J. Chromatogr. Sci., 8, 363 (1970).

443. Hupe, K. P., and Bayer, E., J. Gas Chromatogr., 5, 197 (1967).

444. Hupe, K. P., and Bayer, E., in Advances in Chromatography 1967 (Zlatkis, A., ed.), Preston, Evanston, Ill., 1967.

445. Munk, M. N., Pittsburgh Conference on Analytical Chemistry, Cleveland, 1970, March.

446. Munk, M. N., and Raval, D. N., J. Chromatogr. Sci., 7, 48 (1969).

447. Munk, M. N., Amer. J. Clin. Pathol., 53, 719 (1970).

448. Smuts, T. W., Van Niekerk, F. A., and Pretorius, V., J. Chromatogr. Sci., 7, 127 (1969).

449. Gilbert, T. W., and Dobbs, R. A., Anal. Chem., 45, 1390 (1973).

450. Gurkin, M., and Kallet, E. A., Amer. Lab., 3, 23 (1971).

451. Toporak, M., and Phillip, L. J., J. Chromatogr., 20, 299 (1965).

452. Lawrence, J. F., and Frei, R. W., Anal. Chem., 44, 2046 (1972).

453. Frei, R. W., and Lawrence, J. F., J. Chromatogr., 61, 174 (1971).

454. Frei, R. W., and Lawrence, J. F., J. Chromatogr., 83, 321 (1973).

455. Udenfriend, S., Stein, S., Böhlen, P., Dairman, W., Leimgruber, W., and Weigele, M., Science, 178, 871 (1972).

456. Roth, M., Anal. Chem., 43, 880 (1971).

457. Frei, R. W., Lawrence, F., Hope, J., and Cassidy, R. M., J. Chromatogr. Sci., 12, 40 (1974).

458. Cassidy, R. M., and Frei, R. W., J. Chromatogr., 72, 293 (1972).

459. Thacker, L. H., J. Chromatogr., 73, 117 (1972).

460. Pellizzari, E. D., and Sparacino, C. M., Anal. Chem., 45, 378 (1973).

461. Hatano, H., Yamamoto, Y., Saito, M., Mochida, E., and Watanabe, S., J. Chromatogr., 83, 373 (1973).

462. Blaedel, W. J., and Strohl, J. H., Anal. Chem., 36, 445 (1964).

463. Takemori, Y., and Honda, M., Rev. Polarogr. Jap., 16, 96 (1970).

464. Scarano, E., Bonicelli, M. G., and Forina, M., Anal. Chem., 42, 1470 (1970).

465. Koen, J. G., Huber, J. F. K., Poppe, H., and den Boef, G., J. Chromatogr. Sci., 8, 192 (1970).

466. Joynes, P. L., and Maggs, R. J., J. Chromatogr. Sci., 8, 427 (1970).

467. Pungor, E., and Szepesvary, E., Anal. Chim. Acta, 43, 289 (1968).

468. MacDonald, A., and Duke, P. D., J. Chromatogr., 83, 331 (1973).

469. Schram, E., in Current Status of Liquid Scintillation (Bransome, E. D., Jr., ed.), Grune and Stratton, New York, 1970.

470. Hunt, J. A., Anal. Biochem., 23, 289 (1968).

471. McGuiness, E. T., and Cullen, M. C., J. Chem. Ed., 47, A9 (1970).

472. Sieswerda, G. B., and Polak, H. L., J. Radioanal. Chem., 11, 49 (1972).

473. Schutte, L., J. Chromatogr., 72, 303 (1972).

474. Schutte, L., and Koenders, E. B., J. Chromatogr., 76, 13 (1973).

475. Van Urk-Schoen, A. M., and Huber, J. F. K., Anal. Chim. Acta, 52, 519 (1970).

476. James, A. T., Martin, A. J. P., and Randall, S. S., Biochem. J., 49, 293 (1951).

477. Jackson, A., J. Chem. Ed., 42, 447 (1965).

478. Pecsok, R. L., and Saunders, D. L., Anal. Chem., 40, 1756 (1968).

479. Tesarik, K., and Kalab, P., J. Chromatogr., 78, 357 (1973).

480. Johnson, D. C., and Larochelle, J., Talanta, 20, 959 (1973).

481. Takata, Y., and Muto, G., Anal. Chem., 45, 1864 (1973).

482. Haderka, S., J. Chromatogr., 52, 213 (1970).

483. Haderka, S., J. Chromatogr., 54, 357 (1971).

484. Haderka, S., J. Chromatogr., 57, 181 (1971).

485. Vespalec, R., and Hana, K., J. Chromatogr., 65, 53 (1972).

486. Quillet, R., J. Chromatogr. Sci., 8, 405 (1970).

487. Sternberg, J. C., and Carson, L. M., J. Chromatogr., 2, 53 (1959).

488. Bertin, D., and Grosjean, M., Bull. Soc. Chim. Fr., 1074 (1962).

489. Ford, D. L., and Kennard, W., Oil Colour Chem. Assoc., 49, 299 (1966).

490. Lawrence, J. G., and Scott., R. P. W., Anal. Chem., 39, 831 (1967).

491. Kissinger, P. T., Refshauge, C., Dreiling, R., and Adams, R. N., Anal. Lett., 6, 465 (1973).

492. Yoza, N., and Ohashi, S., Anal. Lett., 6, 595 (1973).

493. Politzer, I. R., Griffin, G. W., Dowty, B. J., and Laseter, J. L., Anal. Lett., 6, 539 (1973).

494. Lovins, R. E., Craig, J., Fairwell, T., and McKinney, C., Anal. Biochem., 47, 539 (1972).

495. Lovins, R. E., Ellis, S. R., Tolbert, G. D., and McKinney, C., Anal. Chem., 45, 1553 (1973).

496. Rao, G. H. R., and Anders, M. W., Fed. Proc., Fed. Amer. Soc., Exp. Biol., 31, 540 (1972).

497. Rao, G. H. R., and Anders, M. W., J. Chromatogr., 84, 402 (1973).

498. Schulten, H. R., and Beckey, H. D., J. Chromatogr., 83, 315 (1973).

499. Waters Associates, Bull., AN 71-103, 1971.

500. Henry, R. A., The 8th Intern. Chromatogr. Symp., Montreaux, 1972.

501. De Stefano, J. J., Diss. Abstr. Int. B, 33, 2976 (1972).

502. Heitz, W., Bomer, B., and Ullner, J., Makromol. Chem., 121, 102 (1969).

503. Mulder, J. L., and Buytenhuys, I. A., J. Chromatogr., 51, 459 (1970).

504. Rickett, F. E., J. Chromatogr., 66, 356 (1972).

505. Stillman, R., and Ma, T. S., Microchim. Acta, 1973, 491.

506. Wolf, J. P., III, Anal. Chem., 45, 1248 (1973).

507. Huber, J. F. K., Van Urk-Schoen, A. M., and Sieswerda, G. B., Z. Anal. Chem., 264, 257 (1973).

508. Huber, J. F. K., van der Linden, R., Ecker, E., and Oreans, M., J. Chromatogr., 83, 267 (1973).

509. Baker, D. R., Henry, R. A., Williams, R. C., Hudson, D. R., and Parris, N. A., J. Chromatogr., 83, 233 (1973).

510. Chang, S. S., in Proc. 3rd Nordic Aroma Symp., Hämeenlinna, June, 1972, p. 203.

511. Burtis, C. A., Goldstein, G., and Scott, C. D., Clin. Chem., 16, 201 (1970).

512. Woodward, R. B., Pure Appl. Chem., 33, 145 (1973).

513. Waters Associates, Bull., AN 72-118, 1972.

514. Schreiber, J., Chimia, 25, 405 (1971).

515. Waters Associates, Bull., AN 72-120, 1972.

516. du Pont LC Technical Bull., 73-2, 1973.

517. Waters Associates, Bull., PB 71-203, 1971.

518. Waters Associates, Bull., PL-1001, 1971.

519. Waters Associates, Bull., PB 69-203, 1969.

520. Waters Associates, Bull., AN 72-123, 1972.

521. Biesenberg, J. A., Tan, M., Duvdevani, I., and Maurer, T., J. Polym. Sci., B9, 429 (1971).

522. Waters, J. L., J. Polym. Sci., Part A2, 8, 411 (1970).

523. Waters, J. L., J. Chromatogr. Sci., 9, 428 (1971).

524. Waters, J. L., J. Chromatogr., 55, 213 (1971).

525. Bombaugh, K. J., Dark, W. A., and Levangie, R. F., J. Chromatogr. Sci., 7, 42 (1969).

526. Bombaugh, K. J., and Levangie, R. F., J. Chromatogr. Sci., 8, 560 (1970).

527. Bombaugh, K. J., J. Chromatogr., 53, 27 (1970).

528. Koreeda, M., Weiss, G., and Nakanishi, K., J. Am. Chem. Soc., 95, 239 (1973).

529. Waters Associates, Bull., AN 125, 1973.

530. Nakamura, S., Ishiguro, S., Yamada, T., and Moriizumi, S., J. Chromatogr., 83, 279 (1973).

531. Janák, J., J. Chromatogr., 78, 117 (1973).

532. Beckman Instruments, Inc., Bull., Model 121, 1973.

533. Technicon Instruments, Co., TMS Analyzer, Bull., 1973.

534. LKB Instruments, Model 3201, Bull., 1973.

535. Perkin Elmer, Model KLA-5, Bull., 1973.

536. Durrum Instrument, Model D-500, Bull., 1973.

537. AMINCO, Aminanalyzer, Bull., 1973.

538. Ertingshausen, G., Adler, H. J., and Reichler, A. S., J. Chromatogr., 42, 355 (1969).

539. Odell, V., Wegener, L., Peczon, B., and Hudson, B. G., J. Chromatogr., 88, 245 (1974).

540. Bellair, T., Anal. Biochem., 54, 578 (1973).

541. Barlow, G. B., and Miles, R., J. Chromatogr., 85, 140 (1973).

542. Carisano, A., Riva, M., and Bonecchi, J., J. Chromatogr., 53, 517 (1970).

543. du Pont LC Product Bull., 830PB3, 1971.

544. Anon., Chem. Eng. News, 1973, March 26.

545. Uziel, M., and Koh, C. K., J. Chromatogr., 59, 188 (1971).

546. Marsden, N. V. B., Öberg, P. A., and Persson, A., J. Chromatogr., 69, 385 (1972).

547. du Pont LC Lab Dialog, 73-3, 1973.

548. Gill, J. M., J. Chromatogr. Sci., 10, 1 (1972).

549. Derge, K., J. Chromatogr., 77, 75 (1973).

550. Gill, J. M., and Henselmann, J., Chromatographia, 5, 108 (1972).

551. Derge, K., Chem. Ztg., 95, 147 (1971).

552. Hubbard, J., Dupre, G. D., and Gill, J. M., Amer. Lab., 1970, Sept.

553. Dupre, G. D., Gill, J. M., and Hubbard, J., Amer. Lab., 1971, Febr.

554. Hettinger, J. D., Hubbard, J., Gill, J. M., and Miller, L. A., J. Chromatogr. Sci., 9, 710 (1971).

555. Gill, J. M., J. Chromatogr. Sci., 7, 731 (1969).

556. Back, H. L., Buttery, P. J., and Gregson, K., J. Chromatogr., 68, 103 (1972).

557. Spitz, H. D., Henyon, G., and Sivertson, J. N., J. Chromatogr., 68, 111 (1972).

558. Fishman, M. L., Landgraff, L. M., and Burdick, D., J. Chromatogr., 86, 37 (1973).

559. Baumann, F., in Basic Liquid Chromatography, Varian, Walnut Creek, 1972, p. 8-11.

560. Chilcote, D. D., and Mrochek, J., Clin. Chem., 17, 751 (1971).

561. Chilcote, D. D., and Scott, C. D., Chem. Instrum., 3, 113 (1971).

562. Scott, C. D., Chilcote, D. D., and Pitt, W. W., Jr., Clin. Chem., 16, 637 (1970).

563. Cerimele, B. J., Clapp, D. C., Cokinos, G. C., Obermeyer, B. D., and Wolen, R. L., Clin. Chem., 18, 744 (1972).

564. Vestergaard, P., and Vedso, S., J. Chromatogr., 19, 512 (1965).

565. Vestergaard, P., Hemmingsen, L., and Hansen, P. W., J. Chromatogr., 40, 16 (1969).

566. Vestergaard, P., Clin. Chem., 16, 651 (1970).

567. Barth, H., Dallmeier, E., Courtois, G., Keller, H. E., and Karger, B. L., J. Chromatogr., 83, 289 (1973).

568. Gouw, T. H., and Jentoft, R. E., in Guide to Modern Methods of Instrumental Analysis (Gouw, T. H., ed.), Wiley (Interscience), New York, 1972.

569. Perry, S. G., Amos, R., and Brewer, P. T., Practical Liquid Chromatography, Plenum, London, 1972.

570. Brown, P. R., High Pressure Liquid Chromatography, Biochemical and Biomedical Applications, Academic Press, New York, 1973.

571. Hatano, H., High Speed Liquid Chromatography, Nankodo Book, Tokyo, 1973.

572. Hatano, H., New Liquid Chromatography, Nankodo Book, Tokyo, 1969.

573. Martin, M., and Guiochon, G., Bull. Soc. Chim. Fr., 1973, 161.

574. Carey, M. A., and Persinger, H. E., J. Chromatogr. Sci., 10, 537 (1972).

575. du Pont LC Methods, 820M8, 1970.

576. Applied Science Lab., Techn. Bull., No. 30, 1972.

577. Reeve Angel, Pellosil Bull., 1972.

578. Chromatronix, LC Appl. No. 14, 1973.

579. Williams, R. C., Baker, D. R., Larmann, J. P., and Hudson, D. R., Intern. Lab., 1973, Nov./Dec.

580. Papa, L. J., and Turner, L. P., J. Chromatogr. Sci., 10, 747 (1972).

581. Hegenauer, J., and Saltman, P., J. Chromatogr., 74, 133 (1972).

582. Kaiser, U. J., Chromatographia, 6, 387 (1973).

583. Churacek, J., and Jandera, P., J. Chromatogr., 53, 69 (1970).

584. Alibert, G., J. Chromatogr., 80, 173 (1973).

585. Hövermann, W., Rapp, A., and Ziegler, A., Chromatographia, 6, 317 (1973).

586. Aurenge, J., J. Chromatogr., 84, 285 (1973).

587. Skelly, N. E., and Crummett, W. B., J. Chromatogr., 55, 309 (1971).

588. Schmit, J. A., and Henry, R. A., Chromatographia, 3, 497 (1970).

589. Longbottom, J. E., Anal. Chem., 44, 418 (1972).

590. Stahl, K. W., Schäfer, G., and Lamprecht, W., J. Chromatogr. Sci., 10, 95 (1972).

591. Prior, R. L., Grunes, D. L., Patterson, R. P., Smith, F. W., Mayland, H. F., and Visek, J., J. Agr. Food Chem., 21, 73 (1973).

592. Katz, S., and Pitt, W. W., Jr., Anal. Lett., 5, 177 (1972).

593. Popl, M., Dolansky, V., and Mostecky, J., J. Chromatogr., 74, 51 (1972).

594. Schmit, J. A., in Modern Practice of Liquid Chromatography (Kirkland, J. J., ed.), Wiley (Interscience), New York, 1971, p. 413.

595. Wu, C. Y., and Siggia, S., Anal. Chem., 44, 1499 (1972).

596. du Pont LC Methods, 820M5, 1970.

597. Talley, C. P., Anal. Chem., 43, 1512 (1971).

598. Vivilecchia, R., Thiebaud, M., and Frei, R. W., J. Chromatogr. Sci., 10, 411 (1972).

599. Waters Associates, Appl. Highlights, No. 4, 1973.

600. Wu, C. Y., Ph.D. thesis, University of Massachusetts, 1973.

601. Wu, C. Y., Siggia, S., Robinson, T., and Waskiewicz, R. D., Anal. Chim. Acta, 63, 393 (1973).

602. Knox, J. H., and Jurand, J., J. Chromatogr., 82, 398 (1973).

603. Knox, J. H., and Jurand, J., J. Chromatogr., 87, 95 (1973).

604. Jolliffe, G. H., and Shellard, E. J., J. Chromatogr., 81, 150 (1973).

605. Heacock, R. A., Langille, K. R., MacNeil, J. D., and Frei, R. W., J. Chromatogr., 77, 425 (1973).

606. Rajcsanyi, P. M., and Rajcsanyi, E., In press.

607. Stutz, M. H., and Sass, S., Anal. Chem., 45, 2134 (1973).

608. Inglis, A. S., and Nicholls, P. W., Anal. Biochem., 24, 209 (1968).

609. Benson, J. R., Amer. Lab., 1972, Oct.

610. Ellis, J. P., Jr., and Garcia, J. B., Jr., J. Chromatogr., 87, 419 (1973).

611. Ellis, J. P., Jr., and Garcia, J. B., Jr., J. Chromatogr., 59, 321 (1971).

612. Liao, T. H., Robinson, G. W., and Salnikow, J., Anal. Chem., 45, 2286 (1973).

613. Weigele, M., De Bernardo, S., Tengi, J., and Leimgruber, W., J. Amer. Chem. Soc., 94, 5927 (1972).

614. Weigele, M., De Bernardo, S., and Leimgruber, W., Biochem. Biophys. Res. Commun., 50, 352 (1973).

615. Stein, S., Böhlen, P., Imai, K., Stone, J., and Udenfriend, S., Fluorescence News, 1973, April.

616. Stein, S., Böhlen, P., Stone, J., Dairman, W., and Udenfriend, S., Arch. Biochem. Biophys., 155, 203 (1973).

617. Felix, A. M., and Terkelsen, G., Anal. Biochem., 56, 610 (1973).

618. Felix, A. M., and Terkelsen, G., Arch. Biochem. Biophys., 157, 177 (1973).

619. Georgiadis, A. G., and Loffey, J. W., Anal. Biochem., 56, 121 (1973).

620. Roth, M., and Hampai, A., J. Chromatogr., 83, 353 (1973).

621. Reeve Angel, Pellionex SCX Bull., 1973.

622. Frank, G., and Strubert, W., Chromatographia., 6, 522 (1973).

623. Graffeo, A. P., Haag, A., and Karger, B. L., Anal. Lett., 6, 505 (1973).

624. Billiet, H. A., Chromatographia, 5, 275 (1972).

625. Williams, A. D., Freeman, D. E., and Florsheim, W. H., J. Chromatogr. Sci., 9, 619 (1971).

626. Chromatronix, LC Appl. No. 11, 1973.

627. Sleight, R. B., Chromatographia, 6, 3 (1973).

628. Varian Instrument Appl., No. 4, 1972.

629. Reeve Angel, Pellosil Bull., 1973.

630. Randau, D., and Bayer, H., J. Chromatogr., 66, 382 (1972).

631. Majors, R. E., J. Chromatogr. Sci., 8, 338 (1970).

632. Reeve Angel, Pellumina Bull., 1973.

633. Reeve Angel, AL-Pellionex WAX Bull., 1973.

634. Oster, H., Van Damme, S., and Ecker, E., Chromatographia, 4, 209 (1971).

635. Amos, R., and Perry, S. G., J. Chromatogr., 83, 245 (1973).

636. du Pont LC Methods, 820M6, 1970.

637. Micro Pack, Varian, Bull., 1972.

638. Veening, H., Pitt, W. W., Jr., and Jones, G., Jr., J. Chromatogr., 90, 129 (1974).

639. Stevenson, R. L., J. Chromatogr. Sci., 9, 251 (1971).

640. Waters Associates, Bull., AN 71-108, 1971.

641. Jentoft, R. E., and Gouw, T. H., Anal. Chem., 40, 1787 (1968).

642. Karger, B. L., Martin, M., Loheac, J., and Guiochon, G., Anal. Chem., 45, 496 (1973).

643. Waters Associates, Bull., AN 72-119, 1972.

644. Waters Associates, Bull., AN 72-121, 1972.

645. Perkin Elmer, Sil-X, Bull., 1972.

646. Popl, M., Dolansky, V., and Mostecky, J., J. Chromatogr., 59, 329 (1971).

647. Reeve Angel, Pellumina Bull. 2, 1973.

648. Karr, C., Childers, E. E., Warner, W. C., and Estep, P. E., Anal. Chem., 36, 2105 (1964).

649. Martinu, V., and Janak, J., J. Chromatogr., 65, 477 (1972).

650. Strubert, W., Chromatographia, 6, 205 (1973).

651. Klimisch, H. J., Anal. Chem., 45, 1960 (1973).

652. Klimisch, H. J., J. Chromatogr., 83, 11 (1973).

653. Ives, N. F., and Giuffrida, L., J. Assoc. Off. Anal. Chem., 55, 757 (1972).

654. Loheac, J., Martin, M., and Guiochon, G., Analusis, 2, 168 (1973).

655. Popl, M., Mostecky, J., Dolansky, V., and Kuras, M., Anal. Chem., 43, 518 (1971).

656. du Pont LC Methods, 820M4, 1970.

657. du Pont LC Methods, 820M2, 1969.

658. Telepchak, M. J., Chromatographia, 6, 234 (1973).

659. Chromatronix, LC Appl. Bull., No. 15, 1973.

660. du Pont LC Product Bull., 830PB1, 1971.

661. Chromatechnology, No. 2, 1973.

662. Schmit, J. A., Henry, R. A., Williams, R. C., and Dieckman, J. F., J. Chromatogr. Sci., 9, 645 (1971).

663. Vaughan, C. G., Wheals, B. B., and Whitehouse, M. J., J. Chromatogr., 78, 203 (1973).

664. Sleight, R. B., J. Chromatogr., 83, 31 (1973).

665. Ledford, C. J., Morie, G. P., and Glover, C. A., Tobacco Sci., 14, 158 (1970).

666. McKay, J. F., and Latham, D. R., Anal. Chem., 45, 1050 (1973).

667. Waters Associates, Bull., AN 71-102A, 1971.

668. Kirkland, J. J., in Modern Practice of Liquid Chromatography (Kirkland, J. J., ed.), Wiley (Interscience), New York, 1971, p. 189.

669. Stehl, R. H., Anal. Chem., 42, 1802 (1970).

670. Samuelson, O., in Ion Exchange, Vol. 2 (Marinsky, J. A., ed.), Marcel Dekker, New York, 1969.

671. Mopper, K., and Degens, E. T., Anal. Biochem., 45, 147 (1972).

672. Otterg, E., Popplewell, J. A., and Taylor, L., J. Chromatogr., 49, 462 (1970).

673. Wu, C. M., Hudson, J. S., and McCready, R. M., Carbohydr. Res., 19, 259 (1971).

674. Kesler, R. B., Anal. Chem., 39, 1416 (1967).

675. Martinsson, E., and Samuelson, O., J. Chromatogr., 50, 420 (1970).

676. Pesek, J. J., and Frost, J. H., Anal. Chem., 45, 1762 (1973).

677. Schneider, J., and Lee, Y. C., Carbohydr. Res., 30, 405 (1973).

678. Watanabe, S., Rokushika, S., Murakami, F., Aoshima, H., and Hatano, H., Anal. Lett., 6, 369 (1973).

679. Hobbs, J. S., and Lawrence, J. G., J. Chromatogr., 72, 311 (1972).

680. Scott, C. D., Attril, J. E., and Anderson, N. G., Proc. Soc. Exptl. Biol. Med., 125, 181 (1967).

681. Scott, C. D., Clin. Chem., 14, 521 (1968).

682. Scott, C. D., Chilcote, D. D., Katz, S., and Pitt, W. W., Jr., J. Chromatogr. Sci., 11, 96 (1973).

683. Scott, C. D., and Pitt, W. W., Jr., J. Chromatogr. Sci., 10, 740 (1972).

684. Scott, C. D., Chilcote, D. D., and Lee, N. L., Anal. Chem., 44, 85 (1972).

685. Chilcote, D. D., and Mrochek, J., Anal. Lett., 6, 531 (1973).

686. Chilcote, D. D., and Mrochek, J., Clin. Chem., 17, 778 (1971).

687. Jolley, R. L., and Scott, C. D., Clin. Chem., 16, 687 (1970).

688. Jolley, R. L., and Freeman, M. L., Clin. Chem., 14, 538 (1968).

689. Jolley, R. L., Warren, K. S., Jainchill, J. L., and Freeman, M. L., Amer. J. Clin. Pathol., 53, 793 (1970).

690. Katz, S., and Burtis, C. A., J. Chromatogr., 40, 270 (1969).

691. Katz, S., Dinsmore, S. R., and Pitt, W. W., Jr., Clin. Chem., 17, 731 (1971).

692. Katz, S., Pitt, W. W., Jr., and Jones, G., Clin. Chem., 19, 817 (1973).

693. Mrochek, J. E., Butts, C. W., Rainey, W. T., Jr., and Burtis, C. A., Clin. Chem., 17, 72 (1971).

694. Mrochek, J. E., Dinsmore, S. R., and Ohrt, D. W., Clin. Chem., 19, 927 (1973).

695. Pitt, W. W., Jr., Scott, C. D., Johnson, W. F., and Jones, G., Clin. Chem., 16, 657 (1970).

696. Pitt, W. W., Jr., Scott, C. D., and Jones, G., Clin. Chem., 18, 767 (1972).

697. Rosevear, J. W., Pfaff, K. J., and Moffitt, E. A., Clin. Chem., 17, 721 (1971).

698. Warren, K. S., and Scott, C. D., Clin. Chem., 15, 1147, (1969).

699. Katz, S., and Thacker, L. H., J. Chromatogr., 64, 247 (1972).

700. Vavich, J. M., and Howell, R. R., Clin. Chem., 16, 702 (1970).

701. Young, D. S., Epley, J. A., and Goldman, P., Clin. Chem., 17, 765 (1971).

702. Young, D. S., Clin. Chem., 16, 681 (1970).

703. Young, D. S., Amer. J. Clin. Pathol., 53, 803 (1970).

704. Butts, C. W., and Jolley, R. L., Clin. Chem., 16, 722 (1970).

705. Scott, C. D., and Lee, N. E., J. Chromatogr., 83, 383 (1973).

706. Burtis, C. A., and Warren, K. S., Clin. Chem., 14, 290 (1968).

707. Burtis, C. A., Butts, C. W., Rainey, W. T., and Scott, C. D., Amer. J. Clin. Pathol., 53, 769 (1970).

708. Kelley, W. N., and Wyngaarden, J. B., Clin. Chem., 16, 707 (1970).

709. Kelley, W. N., and Beardmore, T. D., Science, 169, 388 (1970).

710. Beardmore, T. D., and Kelley, W. N., Clin. Chem., 17, 795 (1971).

711. Burtis, C. A., J. Chromatogr., 52, 97 (1970).

712. Burtis, C. A., and Stevenson, R. L., The Separation of Physiological Fluids by High Resolution LC, Varian, Walnut Creek, 1970.

713. Henderson, T. R., and Jones, R. K., Clin. Chem., 16, 697 (1970).

714. Barness, L. A., Morrow, G., Nocho, R. E., and Maresca, R. A., Clin. Chem., 16, 20 (1970).

715. Shargel, L., Koss, R. F., Crain, A. V. R., and Boyle, V. J., J. Pharm. Sci., 62, 1452 (1973).

716. Martin, W. E., and Cohen, H. P., Anal. Biochem., 53, 177 (1973).

717. Tabor, H., Tabor, C. W., and Irreverre, F., Anal. Biochem., 55, 457 (1973).

718. Seiber, J. N., and Hsieh, D. P. H., J. Ass. Off. Anal. Chem., 56, 827 (1973).

719. Paulson, G. D., Jacobsen, A. M., Zaylskie, R. G., and Feil, V. J., J. Agr. Food Chem., 21, 804 (1973).

720. Simkin, A., Clin. Chem., 16, 191 (1970).

721. Bakke, J. E., and Price, C. E., J. Agr. Food Chem., 21, 640 (1973).

722. Kirkland, J. J., Holt, R. F., and Pease, H. L., J. Agr. Food Chem., 21, 368 (1973).

723. Kirkland, J. J., J. Agr. Food Chem., 21, 171 (1973).

724. Krzeminski, L. F., Cox, B. L., and Neff, A. W., Anal. Chem., 44, 126 (1972).

725. Shepherd, J. A., Nibbelink, D. W., and Stegink, L. D., J. Chromatogr., 86, 173 (1973).

726. Vanderkerckhove, P., and Henderickx, H. K., J. Chromatogr., 82, 379 (1973).

727. Vratny, P., and Zbrozek, J., J. Chromatogr., 76, 482 (1973).

728. Hinsvark, O. N., Zazulak, W., and Cohen, A. I., J. Chromatogr. Sci., 10, 379 (1972).

729. Hinsvark, O. N., Zazulak, W., Kraus, P., and Marinello, J. M., in Progress in Analytical Chemistry (Simmons, I. L., and Ewing, G. W., eds.), Vol. 6, Plenum, New York, 1973, p. 179.

730. Anders, M. W., and Lattore, J. P., Pharmacologist, 12, 273 (1970).

731. Anders, M. W., and Lattore, J. P., J. Chromatogr., 55, 409 (1971).

732. Stermitz, F. R., and Thomas, R. D., J. Chromatogr., 77, 431 (1973).

733. Brown, P. R., and Miech, R. P., Anal. Chem., 44, 1072 (1972).

734. Scott, C. D., Adv. Clin. Chem., 15, 1 (1972).

735. Smith, J. B., Mollica, J. A., Govan, H. K., and Nunes, I. M., Intern. Lab., 1972, July/Aug.

736. Gas-Chrom Newsletter, 1972, Sept./Oct.

737. Pye Unicam, Appl. sheet No. 1, 1972.

738. du Pont LC Methods, 820M3, 1969.

739. Henry, R. A., and Schmit, J. A., Chromatographia, 3, 116 (1970).

740. Stevenson, R. L., and Burtis, C. A., J. Chromatogr., 61, 253 (1971).

741. Reeve Angel, Pellionex SAX Bull., 1972.

742. Waters Associates, Appl. Highlights No. 18, 1971.

743. Waters Associates, Bull., DS 019, 1973.

744. Waters Associates, Bull., DS-71-003, 1971.

745. Chromatography Notes, 1971, Aug.

746. Larson, P., Murgia, E., Hsu, T. J., and Walton, H. F., Anal. Chem. 45, 2306 (1973).

747. Murgia, E., Richards, P., and Walton, H. F., J. Chromatogr., 87, 523 (1973).

748. Chromatronix, LC Appl. No. 13, 1973.

749. Waters Associates, Bull., DS 029, 1973.

750. Anders, M. W., and Lattore, J. P., Anal. Chem., 42, 1430 (1970).

751. Chromatronix, LC Appl. No. 10, 1973.

752. Waters Associates, Appl. Highlights No. 5, 1971.

753. Inaba, T., and Brien, J. F., J. Chromatogr., 80, 161 (1973).

754. Evans, J. E., Anal. Chem., 45, 2428 (1973).

755. Gauchel, V. G., Gauchel, F. D., and Birkofer, L., Z. Klin. Chem. Klin. Biochem., 11, 35 (1973).

756. Roos, R. W., J. Pharm. Sci., 61, 1979 (1972).

757. Jane, I., and Wheals, B. B., J. Chromatogr., 84, 181 (1973).

758. Wittwer, J. D., Jr., and Kluckhohn, J. H., J. Chromatogr. Sci., 11, 1 (1973).

759. Cashman, P. J., Thornton, J. I., and Shelman, D. L., Anal. Chem., 45, 7 (1973).

760. Cashman, P. J., and Thornton, J. I., Forensic Sci. Soc. J., 12, 417 (1972).

761. Kram, T. C., J. Pharm. Sci., 61, 254 (1972).

762. Poet, R. B., and Pu, H. H., J. Pharm. Sci., 62, 809 (1973).

763. Beyer, W. F., Anal. Chem., 44, 1312 (1972).

764. Scott, C. G., and Bommer, P., J. Chromatogr. Sci., 8, 446 (1970).

765. Weber, D. J., J. Pharm. Sci., 61, 1797 (1972).

766. Rajcsanyi, P. M., J. Chromatogr. Sci. (In press.)

767. Kirkland, J. J., in Modern Practice of Liquid Chromatography (Kirkland, J. J., ed.), Wiley (Interscience), New York, 1971, p. 197.

768. Chromatronix, LC Appl. No. 4, 1973.

769. Gas-Chrom Newsletter, 1973, Jan./Feb.

770. Waters Associates, Appl. Highlights No. 1, 1971.

771. Mollica, J. A., Padmanabhan, G. R., and Strusz, R., Anal. Chem., 45, 1859 (1973).

772. Brown, P. R., J. Chromatogr., 52, 257 (1970).

773. Scholar, E. M., Brown, P. R., and Parks, J. E., Jr., Cancer Res., 32, 259 (1973).

774. Nelson, J. A., and Parks, J. E., Jr., Cancer Res., 32, 2034 (1973).

775. Nelson, D. J., Bugge, C. J. L., Krasny, H. C., and Zimmerman, T. P., J. Chromatogr., 77, 181 (1973).

776. Rzeszotarski, W. J., and Mauger, A. B., J. Chromatogr., 86, 246 (1973).

777. Bombaugh, K. J., and Levangie, R. F., Separ. Sci., 5, 751 (1970).

778. Butterfield, A. G., Hughes, D. W., and Pound, N. J., Antimicrob. Agents Chemoter., 4, 115 (1973).

779. Bailey, F., and Brittain, P. N., J. Chromatogr., 83, 431 (1973).

780. Muusze, R. G., and Huber, J. F. K., J. Chromatogr., 83, 405 (1973).

781. Mikes, F., Schurig, V., and Gil-Av, E., J. Chromatogr., 83, 91 (1973).

782. Dunham, E. W., and Anders, M. W., Prostaglandins, 4, 85 (1973).

783. Pound, N. J., McGilveray, I. J., and Sears, R. W., J. Chromatogr., 89, 23 (1974).

784. Pound, N. J., and Sears, R. W., Can. J. Pharm. Sci., 8, 84 (1973).

785. Grady, L. T., Hays, S. E., King, R. H., Klein, H. R., Mader, W. J., Wyatt, D. K., and Zimmere, R. O., J. Pharm. Sci., 62, 456 (1973).

786. Skelly, N. E., and Cornier, R. F., J. Ass. Off. Anal. Chem., 54, 551 (1971).

787. Reeve Angel, Progress Report on Pellidon, 1973.

788. Wildanger, W. A., Chromatographia, 6, 381 (1973).

789. Nelson, J. F., J. Chromatogr. Sci., 11, 28 (1973).

790. Palmer, J. K., and List, D. M., J. Agr. Food Chem., 21, 903 (1973).

791. Stahl, E., and Laub, E., Z. Lebensm. Unters. Forsch., 152, 280 (1973).

792. Martin, G. E., Guinand, G. G., and Figert, D. M., J. Agr. Food Chem., 21, 544 (1973).

793. Walradt, J. P., and Shu, C. K., J. Agr. Food Chem., 21, 547 (1973).

794. Schmit, J. A., Williams, R. C., and Henry, R. A., J. Agr. Food Chem., 21, 551 (1973).

795. Stewart, I., and Wheaton, T. A., J. Chromatogr., 55, 325 (1971).

796. Sweeney, J. P., and Marsh, A. C., J. Ass. Off. Anal. Chem., 53, 937 (1971).

797. Van de Weerdhof, K., Wiersum, M. L., and Reissenweber, H., J. Chromatogr., 83, 455 (1973).

798. Cox, G. B., J. Chromatogr., 83, 471 (1973).

799. Hobbs, J. S., and Lawrence, J. G., J. Sci. Food Agr., 23, 45 (1972).

800. Aitzetmüller, K., Fette, Seifen, Anstrichm., 74, 598 (1972).

801. Aitzetmüller, K., Fette, Seifen, Anstrichm., 75, 14 (1973).

802. Aitzetmüller, K., Fette, Seifen, Anstrichm., 75, 256 (1973).

803. Aitzetmüller, K., J. Chromatogr., 73, 248 (1972).

804. Aitzetmüller, K., J. Chromatogr., 71, 355 (1972).

805. Aitzetmüller, K., J. Chromatogr., 83, 461 (1973).

806. Chromatec, Bull., PB-203, 1973.

807. Stevenson, R. L., and Burtis, C. A., Clin. Chem., 17, 114 (1971).

808. Palamand, S. R., and Aldenhoff, F. M., J. Agr. Food Chem., 21, 535 (1973).

809. Molyneux, R. J., and Wong, Y., J. Agr. Food Chem., 21, 531 (1973).

810. Drawert, F., Beier, J., and Merle, W., Chromatographia, 6, 160 (1973).

811. Vanheertum, R., and Verzele, M., J. Inst. Brew., London, 79, 324 (1973).

812. Kleber, W., and Hums, N., Braunwelt, 113, 411 (1973).

813. Hansen, G., and Ramus, S. E., Amer. Soc. Brew. Chem. Proc., 1971, 355.

814. Vanheertum, R., Chromatographia, 6, 390 (1973).

815. Vanheertum, R., Chromatographia, 6, 217 (1973).

816. Siebert, K. J., Tech. Quart., Master Brew. Assoc. Amer., 9, 205 (1972).

817. Frache, R., and Dadone, A., Chromatographia, 5, 581 (1972).

818. Strelow, F. W. E., and Boshoff, M. D., Anal. Chim. Acta, 62, 351 (1972).

819. Seymour, M. D., Sickafoose, J. P., and Fritz, J. S., Anal. Chem., 43, 1734 (1971).

820. Karol, P. J., J. Chromatogr., 79, 287 (1973).

821. Campbell, D. O., and Buxton, S. R., Ind. Eng. Chem. Process. Res. Devel., 9, 89 (1970).

822. Foti, S. C., and Wish, L., J. Chromatogr., 29, 203 (1967).

823. Campbell, D. O., and Ketelle, B. H., Inorg. Nucl. Chem. Lett., 5, 533 (1969).

824. Yamabe, T., and Hayashi, T., J. Chromatogr., 76, 213 (1973).

825. Campbell, D. O., Ind. Eng. Chem. Process. Res. Devel., 9, 95 (1970).

826. Sisson, D. H., Mode, V. A., and Campbell, D. O., J. Chromatogr., 66, 129 (1972).

827. Huber, J. F. K., Kraak, J. C., and Veening, H., Anal. Chem., 44, 1554 (1972).

828. Veening, H., Greenwood, J. M., Shanks, W. H., and Willeford, B. R., Chem. Commun., 1969, 1305.

829. Greenwood, J. M., Veening, H., and Willeford, B. R., J. Organometal Chem., 38, 345 (1972).

830. Garden, S. A., Seyler, R., Veening, H., and Willeford, B. R., J. Organometal Chem., 39, 418 (1973).

831. Veening, H., in Progress in Analytical Chemistry (Simmons, I. L., and Ewig, G. W., eds.), Vol. 6, Plenum, New York, 1973, p. 165.

832. Huber, J. F. K., and Van Urk-Schoen, A. M., Anal. Chim. Acta, 58, 395 (1972).

833. Seymour, M. D., and Fritz, J. S., Anal. Chem., 45, 1394 (1973).

834. Seymour, M. D., Diss. Abstr. Int. B, 33, 4703 (1973).

835. Seymour, M. D., and Fritz, J. S., Anal. Chem., 45, 1632 (1973).

836. Horwitz, E. P., and Bloomquist, C. A. A., J. Chromatogr. Sci., 12, 11 (1974).

837. Evans, W. J., and Hawthorne, M. F., J. Chromatogr., 88, 187 (1974).

838. Dustin, D. F., Evans, W. J., Jones, C. J., Wiersma, R. J., and Hawthorne, M. F., J. Amer. Chem. Soc. (In press.)

839. Dustin, D. F., Dunks, G. B., and Hawthorne, M. F., J. Amer. Chem. Soc., 95, 1109 (1973).

840. Yamamoto, Y., Yamamoto, M., Ebisui, S., Takagi, T., Hashimoto, T., and Izuhara, M., Anal. Lett., 6, 451 (1973).

841. Kawazu, K., and Fritz, J. S., J. Chromatogr., 77, 397 (1973).

842. Burtis, C. A., and Gere, D., Nucleic Acid Constituents by Liquid Chromatography, Varian, Walnut Creek, 1970.

843. Gere, D. R., in Modern Practice of Liquid Chromatography (Kirkland, J. J., ed.), Wiley (Interscience), New York, 1971, p. 417.

844. Rajcsanyi, P. M., Csillag, M., and Rajcsanyi-Kriskovics, E., Separation Purification Meth., 3, 167 (1974).

845. Horvath, C. G., and Lipsky, S. R., Anal. Chem., 41, 1227 (1969).

846. Uziel, M., Koh, C. K., and Cohn, W. E., Anal. Biochem., 25, 77 (1968).

847. Burtis, C. A., Fed. Proc., 29, 726 (1970).

848. Hanson, C. V., Anal. Biochem., 32, 303 (1969).

849. Kennedy, W. P., and Lee, J. C., J. Chromatogr., 51, 203 (1970).

850. Shmukler, H., J. Chromatogr. Sci., 8, 653 (1970).

851. Shmukler, H., J. Chromatogr. Sci., 10, 38 (1972).

852. Shmukler, H., J. Chromatogr. Sci., 10, 137 (1972).

853. Kirkland, J. J., J. Chromatogr. Sci., 8, 72 (1970).

854. Burtis, C. A., Munk, M. N., and McDonald, F. R., Clin. Chem., 16, 667 (1970).

855. Brown, P. R., J. Chromatogr., 57, 383 (1971).

856. du Pont LC Methods, 820M11, 1972.

857. Gabriel, T. F., and Michalewsky, J. E., J. Chromatogr., 67, 309 (1972).

858. Henry, R. A., Schmit, J. A., and Williams, R. C., J. Chromatogr. Sci., 11, 358 (1973).

859. Reeve Angel, AS Pellionex SAX, Bull., 1973.

860. Brown, P. R., Anal. Biochem., 43, 305 (1971).

861. Shimizu, T., Bunseki Kagaku, 18, 632 (1969).

862. Pennington, S. N., Anal. Chem., 43, 1701 (1971).

863. Brooker, G., J. Biol. Chem., 246, 7810 (1971).

864. Brooker, G., in New Assay Methods for Cyclic Nucleotides (Greengard, P., Robinson, G. A., and Paoletti, R., eds.), North Holland, 1972.

865. Brooker, G., Anal. Chem., 42, 1108 (1970).

866. Brooker, G., Advan. Cyclic Nucleotide Res., 2, 111 (1972).

867. Pickett, J. A., J. Chromatogr., 81, 156 (1973).

868. Stahl, K. W., Schlimme, E., and Schäfer, G., J. Chromatogr., 76, 477 (1973).

869. Stahl, K. W., Schlimme, E., and Bojanowski, D., J. Chromatogr., 83, 395 (1973).

870. Virkola, P., J. Chromatogr., 51, 195 (1970).

871. Drobishev, V. I., Mansurova, S. E., and Kulaev, I. S., J. Chromatogr., 69, 317 (1972).

872. Breter, H. J., and Zahn, R. K., Anal. Biochem., 54, 346 (1973).

873. Chromatography Notes, 1973, Jan.

874. Stegman, R. J., Senft, A. W., Brown, P. R., and Parks, R. E., Jr., Biochem. Pharmacol. (In press.)

875. Varian, Series 4000 LC, Bull., 1972.

876. Clifford, A. J., Riumallo, J. A., Baliga, B. S., Munro, H. N., and Brown, P. R., Biochim. Biophys. Acta, 277, 443 (1972).

877. Brown, P. R., Agarwal, R. P., Gell, J., and Parks, R. E., Jr., Comp. Biochem. Physiol., 43B, 891 (1972).

878. Brown, P. R., and Parks, R. E., Jr., Anal. Chem., 45, 948 (1973).

879. Scholar, E. M., Brown, P. R., Parks, R. E., Jr., and Calabresi, P., Blood, 41, 927 (1973).

880. Brown, P. R., Parks, R. E., Jr., and Herod, J., Clin. Chem., 19, 919 (1973).

881. Senft, A. W., Miech, R. P., Brown, P. R., and Senft, D., Int. Parasitol., 2, 249 (1972).

882. Miech, R. P., and Tung, M. C., Biochem. Med., 4, 435 (1970).

883. Scott, R. P. W., Scott, C. G., and Kucera, P., Anal. Chem., 44, 100 (1972).

884. Burtis, C. A., J. Chromatogr., 51, 183 (1970).

885. Chromatronix, LC Appl. No. 5, 1973.

886. Busch, W. E., J. Chromatogr., 37, 518 (1968).

887. Busch, W. E., Borcke, I. M., and Martinez, B., Biochem. Biophys. Acta, 166, 547 (1968).

888. Busch, W. E., Borcke, I. M., Greve, H., and Thorn, W., Z. Physiol. Chem., 349, 801 (1968).

889. Junowicz, E., and Spencer, J. H., J. Chromatogr., 44, 342 (1969).

890. Burtis, C. A., Gere, D., Gill, J. M., and MacDonald, F. R., Chromatographia, 3, 161 (1970).

891. Chromatronix, LC Appl. No. 7, 1973.

892. Chromatronix, LC Appl. No. 12, 1973.

893. Murakami, F., Rokushika, S., and Hatano, H., J. Chromatogr., 53, 584 (1970).

894. Gauchel, G., Gauchel, F. D., Beyermann, K., and Zahn, R. K., Z. Anal. Chem., 259, 183 (1972).

895. Green, J. G., Nunley, C. E., and Anderson, N. G., Natl. Cancer Inst. Monogr., 21, 431 (1966).

896. Anderson, N. G., Green, J. G., Barber, M. L., and Ladd, F. C., Anal. Biochem., 6, 431 (1963).

897. Byrne, P. V., and Chapman, I. V., J. Chromatogr., 88, 190 (1974).

898. Pfadenhauer, E. H., J. Chromatogr., 81, 85 (1973).

899. Brown, P. R., and Parks, R. E., Jr., Pharmacologist, 13, 210 (1971).

900. Burtis, C. A., and Gere, D., in 5th Intern. Symp. on Column Liquid Chromatography (Kovats, E., ed.), Lausanne, 1970, p. 51.

901. Gabriel, T. F., and Michalewsky, J. E., J. Chromatogr., 80, 263 (1973).

902. Cook, A. F., DeCzekala, A., Gabriel, T. F., Harvey, C. L., Holman, M., Michalewsky, J. E., and Nussbaum, A. L., Biochem. Biophys. Acta, 324, 433 (1973).

903. Gabriel, T. F., and Michalewsky, J. E., Intern. Lab., 1973, Nov./Dec.

904. Vlasov, V. V., Grachev, M. A., Komarova, N. I., Kuzmin, S. V., and Menzorova, N. I., Mol. Biol., 6, 809 (1972).

905. Egan, B. Z., Anal. Biochem., 56, 616 (1973).

906. Duch, D. S., Borlowska, I., Stasink, L., and Laskowski, M., Anal. Biochem., 53, 459 (1973).

907. Kelmers, A. D., and Heatherly, D. E., Anal. Biochem., 44, 486 (1971).

908. Schmit, J. A., in Modern Practice of Liquid Chromatography (Kirkland, J. J., ed.), Wiley (Interscience), New York, 1971, p. 400.

909. Waters Associates, Bull., AN 72-115, 1972.

910. Pye Unicam, Appl. sheet No. 3, 1972.

911. du Pont LC Methods, 820M12, 1972.

912. du Pont LC Methods, 820M7, 1970.

913. Schmit, J. A., in Modern Practice of Liquid Chromatography (Kirkland, J. J., ed.), Wiley (Interscience), New York, 1971, p. 405.

914. du Pont LC Appl. Lab. Report, 72-07, 1972.

915. Kirkland, J. J., Anal. Chem., 41, 219 (1969).

916. Eisenbeiss, F., and Sieper, H., J. Chromatogr., 83, 439 (1973).

917. Chromatronix, LC Appl. No. 3, 1973.

918. Chromatronix, LC Appl. No. 9, 1973.

919. Varian, Variscan Bull., 1972.

920. Porcaro, P. J., and Shubiak, P., Anal. Chem., 44, 1865 (1972).

921. Henry, R. A., Schmit, J. A., Dieckman, J. F., and Murphey, F. J., Anal. Chem., 43, 1053 (1971).

922. Machin, A. F., Pestic. Sci., 4, 425 (1973).

923. Pye Unicam, Appl. sheet, No. 2, 1972.

924. Column, No. 15, 1972.

925. Reeve Angel, Pellidon Bull., 1972.

926. Fritz, J. S., and Willis, R. B., J. Chromatogr., 79, 107 (1973).

927. Siemens, S200P Bull., 1972.

928. Kramer, U., and Bhatia, K., J. Chromatogr., 88, 348 (1974).

929. Bhatia, K., Anal. Chem., 45, 1344 (1973).

930. Gudzinowicz, B. J., and Alden, K., J. Chromatogr. Sci., 9, 65 (1971).

931. Little, J. N., Waters, J. L., Bombaugh, K. J., and Pauplis, N. J., J. Polym. Sci., A2-7, 1775 (1969).

932. Little, J. N., Waters, J. L., Bombaugh, K. J., and Pauplis, W. J., J. Chromatogr. Sci., 9, 341 (1971).

933. Heitz, W., Klatyk, K., Kraffczyk, F., Pfitzner, K., and Randau, D., J. Chromatogr. Sci., 9, 521 (1971).

934. Little, J. N., Waters, J. L., Bombaugh, K. J., and Pauplis, W. J., Separation Sci., 5, 765 (1970).

935. Billmeyer, F. W., and Altgelt, K. H., Separation Sci., 5, 393 (1970).

936. LePage, M., Beau, R., and De Vries, A. J., J. Polym. Sci., Part C, 21, 119 (1968).

937. Chitumbo, K., and Brown, W., J. Chromatogr., 80, 187 (1973).

938. Bombaugh, K. J., and Levangie, R. F., Anal. Chem., 41, 1357 (1969).

939. Bombaugh, K. J., King, R. N., and Cohen, A. J., J. Chromatogr., 35, 108 (1968).

940. Bombaugh, K. J., Dark, W. A., and Levangie, R. F., Z. Anal. Chem., 236, 443 (1968).

941. Bombaugh, K. J., in Modern Practice of Liquid Chromatography (Kirkland, J. J., ed.), Wiley (Interscience), New York, 1971, p. 371.

942. Cantow, M. J. R., and Johnson, J. F., J. Polym. Sci., 5A-1, 2835 (1967).

943. Quinn, E. J., Osterdout, H. W., Heckles, J. S., and Ziegler, D. C., Anal. Chem., 40, 547 (1968).

944. Ede, P. S., J. Chromatogr. Sci., 9, 275 (1971).

945. Hendrickson, J. F., Anal. Chem., 40, 49 (1968).

946. Hendrickson, J. F., and Moore, J. C., J. Polym. Sci., A4, 167 (1966).

947. Huber, J. F. K., Kolder, F. F. M., and Miller, J. M., Anal. Chem., 44, 105 (1972).

948. Siggia, S., and Dishman, R. A., Anal. Chem., 42, 1223 (1970).

949. Huber, J. F. K., Hulsman, J. A. R. J., and Meijers, C. A. M., J. Chromatogr., 62, 79 (1971).

950. Hulsman, J. A. R. J., Ph.D. thesis, University of Amsterdam, 1969.

951. Huber, J. F. K., Hulsman, J. A. R. J., and Meijers, C. A. M., Z. Anal. Chem., 261, 347 (1972).

952. Meijers, C. A. M., Ph.D. thesis, University of Amsterdam, 1971.

953. Henry, R. A., Schmit, J. A., and Dieckman, J. F., J. Chromatogr. Sci., 9, 513 (1971).

954. Dolphin, R. J., J. Chromatogr., 83, 421 (1973).

955. Chromatography Notes, 1971, Dec.

956. Butterfield, A. G., Lodge, B. A., and Pound, N. J., J. Chromatogr. Sci., 11, 401 (1973).

957. Hesse, C., and Hövermann, W., Chromatographia, 6, 345 (1973).

958. Touchstone, J. C., and Wortmann, W., J. Chromatogr., 76, 244 (1973).

959. Wortmann, W., Schnabel, C., and Touchstone, J. C., J. Chromatogr., 84, 396 (1973).

960. Cavina, G., Moretti, G., and Cantafora, A., J. Chromatogr., 80, 89 (1973).

961. Cavina, G., Moretti, G., Mollica, A., and Siniscalchi, P., J. Chromatogr., 44, 493 (1969).

962. Cavina, G., Mollica, A., Moretti, G., and Antonini, R., J. Chromatogr., 60, 179 (1971).

963. Gas-Chrom Newslett., 1972/2.

964. Waters Associates, Bull., AN 73-122, 1973.

965. Waters Associates, Bull., AN 124, 1973.

966. Lötscher, K., and Kern, H., Chimia, 27, 348 (1973).

967. Majors, R. E., Varian Instrument Appl., 1973, No. 4.

968. Fitzpatrick, F. A., Siggia, S., and Dingman, J., Anal. Chem., 44, 2211 (1972).

969. Fitzpatrick, F. A., Diss. Abstr. Int. B, 33, 5705 (1973).

970. Fitzpatrick, F. A., and Siggia, S., Anal. Chem., 45, 2310 (1973).

971. Evans, F. J., J. Chromatogr., 88, 411 (1974).

972. Waters Associates, Bull., AN 71-114, 1971.

973. Kram, T. C., FDA By-Lines, 1, 290 (1971).

974. Bailey, F., and Brittain, P. N., J. Pharm. Pharm., 24, 425 (1972).

975. Fitzpatrick, F. A., Clin. Chem., 19, 1293 (1973).

976. Landgraf, W. C., and Jennings, E. C., J. Pharm. Sci., 62, 278 (1973).

977. Mollica, J. A., and Strusz, R. F., J. Pharm. Sci., 61, 444 (1972).

978. Karger, B. L., and Berry, L. V., Clin. Chem., 17, 757 (1971).

979. Unger, K., and Nyamah, D., Chromatographia, 7, 63 (1974).

980. Williams, R. C., Baker, D. R., and Schmit, J. A., J. Chromatogr. Sci., 11, 618 (1973).

981. du Pont LC Methods, 820M10, 1972.

982. du Pont LC Lab. Report, 72-09, 1972.

983. Williams, R. C., Schmit, J. A., and Henry, R. A., J. Chromatogr. Sci., 10, 494 (1972).

984. Krol, G. J., J. Chromatogr., 74, 43 (1972).

985. Khym, J. X., Jolley, R. L., and Scott, C. D., Cereal Sci. Today, 15, 44 (1970).

986. Stevenson, R. L., in Basic Liquid Chromatography, Varian, Walnut Creek, 1972, pp. 9-24.

987. Chromatronix, LC Appl. No. 1, 1973.

988. Waters Associates, Bull., AN 71-111, 1971.

989. Thompson, J. N., Erdody, P., and Maxwell, W. B., Anal. Biochem., 50, 267 (1972).

990. Van Niekerk, P. J., Anal. Biochem., 52, 533 (1973).

991. Vecchi, M., Vesely, J., and Oesterhelt, G., J. Chromatogr., 83, 447 (1973).

992. Perkin Elmer, Bull., 1969.

993. du Pont LC Methods, 820M1, 1969.

994. Chromatography Notes, 1971, Jan.

995. Waters Associates, Bull., AN 71-110, 1971.

996. Karger, B. L., J. Chromatogr. Sci., 9, 325 (1971).

997. Fishbein, L., and Albro, P. W., J. Chromatogr., 70, 365 (1972).

998. Rouser, G., J. Chromatogr. Sci., 11, 60 (1973).

999. Stolyhwo, A., and Privett, O. S., J. Chromatogr. Sci., 11, 20 (1973).

1000. Lawrence, J. G., J. Chromatogr., 84, 299 (1973).

1001. Montet, J. C., Amic, J., and Hauton, J. C., Bull. Soc. Chim. Biol., 52, 831 (1970).

1002. Amic, J., Lairon, D., and Hauton, J. C., Clin. Chim. Acta, 40, 107 (1972).

1003. Lairon, D., Amic, J., Lafont, H., Nalbone, G., Domingo, N., and Hauton, J., J. Chromatogr., 88, 183 (1974).

1004. Lairon, D., Ph.D. thesis, University of Provence, 1972.

1005. Katz, S., Pitt, W. W., Jr., Scott, C. D., and Rosen, A. A., Water Res., 6, 1 (1972).

1006. Chapman, J. N., and Beard, H. R., Anal. Chem., 45, 2268 (1973).

1007. Doali, J. O., and Juhasz, A. A., J. Chromatogr. Sci., 12, 51 (1974).

1008. Deelder, R. S., and Hendricks, P. J. H., J. Chromatogr., 83, 343 (1973).

LIST OF SYMBOLS

K	Partition (distribution) coefficient
V_R	Retention volume
V_s	Stationary phase volume in the column
V_m	Interstitial volume of the column
V_N	Net retention volume
V_g	Specific retention volume
w_s	Weight of stationary phase
k	Capacity factor (partition ratio)
β	Phase ratio
μ	Chemical potential
γ	Solute activity coefficient
t_R	Retention time
t_0	Average residence of the mobile phase in the column
α	Relative retention
N	Plate number
H	Plate height
u	Linear velocity
ΔP	Pressure drop
d_p	Particle diameter
S	Separation factor

C_0	Constant
h	Reduced plate height
v	Reduced linear velocity
K_1	Permeability
C_m	Mass transfer resistance constant
d_c	Column diameter
L	Column length
η	Viscosity of the mobile phase
f_1	Total porosity in the column
K_2	Column permeability parameter
R	Resolution
y	Constant
ρ_s	Stationary phase density

AUTHOR INDEX

Numbers in parentheses are reference numbers and indicate that an author's work is referred to although his name is not cited in the text. Numbers underlined show the page on which the complete reference is cited.

A

Abel, E. W., 36(284), 137
Abrahams, L., 28(198), 30(198), 75(198), 77(198), 108(198), 133
Adams, R. N., 57(491), 76(491), 145
Adler, H. J., 60(538), 71(538), 147
Adler, R. J., 31(235), 135
Adrian, P., 37(301), 138
Agarwal, R. P., 105(877), 162
Aitzetmüller, K., 100, 159
Albro, P. W., 121, 167
Alden, K., 111(930), 164
Aldenhof, J. M., 100, 159
Alderlieste, E. T., 4(21), 114(21), 126
Ali, S. L., 23(148), 131
Alibert, G., 66(584), 149
Allington, W. B., 23(174), 64(174), 132
Altgelt, K. H., 111(935), 125, 165
Amic, J., 122(1001-1003), 167
Amos, R., 12(79), 32(79), 63(569), 75(635), 78(79), 128, 149, 151
Anders, M. W., 57(496, 497), 89(497), 90, 92, 96, 146, 156, 158
Anderson, N. G., 85(680), 105(895, 896), 153, 163
Anderson, R. A., 1(13), 13(13), 125
Ando, T., 7(40-42), 127
Angl, J. M., 36(293), 137
Antonini, R., 115(962), 166
Aoshima, H., 83(678), 153
Arikawa, Y., 23(126), 130
Asshauer, J., 17(98), 39(98), 64(98), 129
Attril, J. E., 85(680), 153
Aue, W., 36(292, 293), 137

Aurenge, J., 67(586), 121(586), 149

B

Back, H. L., 61(556), 148
Bailey, F., 96(779), 118, 158, 166
Baker, D. R., 58, 65(579), 74(579), 97(579), 108(509), 118(980), 146, 149, 167
Bakke, J. E., 89(721), 155
Bakken, M., 48, 142
Baliga, B. S., 105(876), 162
Banner, B., 36(295), 137
Barber, M. L., 105(896), 163
Barlow, G. B., 60(541), 147
Barness, L. A., 88, 155
Barrall, E. M., 32(252), 136
Barth, H., 31, 61, 135, 149
Baumann, F., 31(238), 42(324), 61(559), 135, 139, 148
Bayer, E., 50, 143
Bayer, H., 75(630), 110(630), 151
Beachell, H. C., 10(63, 64), 13, 15, 16, 18, 57, 58(64), 64(94), 76(94), 128, 129
Beard, H. R., 122, 168
Beardmore, T. D., 87, 155
Beau, R., 34(275), 111(936), 136, 165
Beckey, H. D., 57(498), 115, 146
Beier, J., 100(810), 159
Beisel, V., 37(305), 138
Bektesh, S. L., 38(307, 308), 75(307), 138
Bellair, T., 60(540), 147
Benoit, H., 4(20), 126
Benson, J. R., 71(609), 150
Berg, K., 37(302-304), 138
Berry, L. V., 28, 29, 30(215), 118, 134, 166
Bertin, D., 56(488), 145
Beyer, W. F., 95(763), 157
Beyermann, K., 105(894), 163
Bhatia, K., 111(928, 929), 122(929), 164
Bidlingmeyer, B. A., 28, 30(202), 133
Biesenberg, J. A., 43(332), 59(521), 139, 146
Billiet, H. A., 73(624), 151
Billmeyer, F. W., 111(935), 165
Birkofer, L., 93(755), 157
Bitterfield, Z., 23(138), 99(138), 131
Blackburn, D. W. J., 12(73), 31(232), 128, 135
Blaedel, W. J., 53(462), 144
Bloomquist, C. A. A., 102, 160
Blu, G., 9(49), 127
Bly, D. L., 34(270), 136
Bogue, D. C., 1(13), 13(13), 125
Bojanowski, D., 105(869), 162
Bombaugh, K. J., 23(127-129, 154, 156), 27, 28(198), 30(198), 32(243, 245, 246), 34(245-247), 36(128), 42(192), 59(525-527), 74(274), 75(198), 78(198), 96, 108(198), 111(243, 525, 527, 931, 932, 934, 938, 939), 112(940), 113(940, 941), 130, 131, 133, 135, 136, 146, 158, 164, 165
Bomer, B., 57(502), 146
Bommer, P., 95(764), 157
Bonecchi, J., 60(542), 147
Bonicelli, M. G., 53(464), 144
Bonneleycke, B. E., 28(204), 133
Borcke, I. M., 105(887, 888), 162
Borlowska, I., 106(906), 163
Boshoff, M. D., 101(818), 159
Bossart, C. J., 36(289), 137
Boyle, V. J., 88(715), 155
Böhlen, P., 52(455), 72(455, 615, 616), 144, 151
Brdicka, R., 46(379), 141
Breter, H. J., 105, 162
Brewer, P. T., 63(569), 149
Brien, J. F., 93(753), 157
Brittain, P. N., 96(779), 118, 158, 166

Brooker, G., 46(389), 105(863-866), 141, 161
Brown, P. R., 63(570), 91(733), 96(772, 773), 102(570), 105(772, 773, 855, 860, 874, 876-881), 106(899), 149, 156, 157, 161-163
Brown, W., 111(937), 165
Brust, O. E., 37, 138
Bruzzone, A. R., 32(252), 136
Bugge, C. J. L., 96(775), 157
Burdick, D., 61(558), 148
Burke, M. F., 36(287), 137
Burnev, N. P., 49(415), 142
Burtis, C. A., 59, 85, 87, 91, 92(740), 100, 102(842), 103, 105(854, 884, 890), 106(900), 146, 154-156, 159, 161-163
Busch, W. E., 105(886-888), 162
Butterfield, A. G., 29(213), 96, 115(956), 134, 158, 166
Buttery, P. J., 61(556), 148
Butts, C. W., 19(119), 21(119), 85(693, 704, 707), 130, 154, 155
Buxton, S. R., 101(821), 159
Buytenhuys, I. A., 57(503), 146
Bylina, A., 47, 141
Byrne, P. V., 105(897), 163
Byrne, S. H., 26, 44(352), 60(187), 133, 140

C

Cain, J. H., 32(252), 136
Calabresi, P., 105(879), 162
Callmer, K., 47, 141
Campanile, V. A., 48(400), 141
Campbell, D. O., 101, 159
Cantafora, A., 115(960), 166
Cantow, M. J. R., 32(247), 113(942), 135, 165
Carey, M. A., 63(574), 65(574), 149
Carisano, A., 60(542), 147
Carr, D., 45(355), 119(355), 140
Carson, L. M., 56(487), 145
Cashaw, J. L., 50(442), 143
Cashman, P. J., 94(759, 760), 157
Cassidy, R. M., 30, 52, 134, 144
Casto, M. E., 32(248), 135
Cavina, G., 115(960-962), 166
Cerimele, B. J., 61, 148
Chandler, D. C., 23(157), 38, 64(312), 66(312), 132, 138
Chang, S. S., 58, 146
Chang, T. L., 32(249), 135
Chapman, I. V., 105(897), 163
Chapman, J. N., 122, 168
Chernov, A. Z., 16(95), 116(95), 129
Chilcote, D. D., 19(116, 117), 26, 60(189), 61, 85(562, 682, 684-686), 87, 130, 133, 148, 153, 154
Childers, E. E., 1(14), 78(648), 125, 152
Chitumbo, K., 111(937), 165
Churacek, J., 66(583), 149
Clapp, D. C., 61(563), 148
Claxton, G. C., 50(439), 143
Clifford, A. J., 105(876), 162
Cohen, A. I., 90(728), 156
Cohen, A. J., 27(192), 42(192), 111(939), 133, 165
Cohen, H. P., 88(716), 155
Cohn, W. E., 103(846), 161
Cokinos, G. C., 61(563), 148
Coll, H., 49(432), 143
Conlon, R. D., 23(153), 27(194), 43(344, 350), 47(344), 131, 133, 140
Conroe, K., 39(315), 40(320), 111(315), 138
Cook, A. F., 106(902), 163
Cooper, A. R., 23(130), 32(252, 253), 131, 136
Cornier, R. F., 97, 158
Courtois, G., 61(567), 149
Cox, B. L., 90(724), 155
Cox, G. B., 99(798), 159
Craig, J., 57(494), 100(494), 146
Crain, A. V. R., 88(715), 155
Cranston, R. W., 30(228),

[Cranston, R. W.] 60(228), 134
Cropper, F. R., 49(422), 142
Crummett, W. B., 67(587), 149
Csillag, M., 102(844), 161
Culberson, C. F., 23(155), 131
Cullen, M. C., 54, 145

D

Dadone, A., 101(817), 159
Dairman, W., 52(455), 72(455, 616), 144, 151
Dallmeier, E., 31(233), 61(567), 135, 149
Dark, W. A., 32(243), 34(272), 59(525), 111(243, 525), 112(940), 113(940), 135, 136, 146, 165
De Bernardo, S., 72(613, 614), 151
De Clerk, 19(106), 130
De Czekala, A., 106(902), 163
Deelder, R. S., 8(46), 32(46), 35(46), 123, 127, 168
Degens, E. T., 46(381), 82(671), 105, 141, 153
Deininger, G., 9(51), 18(101), 28(208, 209), 29(208), 43(349), 46(349), 48, 127, 129, 134, 140, 142
Delfel, N. E., 26, 60(188), 133
Den Boef, G., 53(465), 144
Denson, K. B., 38(307), 75(307), 138
Derge, K., 61, 148
De Stefano, J. J., 10(63, 64), 13, 15, 16, 18(64), 33(262), 34, 36(294), 57, 58(64), 64(94), 76(94), 107, 128 129, 136, 137, 146
Determann, H., 43(333), 125, 139
De Vries, A. J., 34, 111(936), 136, 165
Dieckman, J. F., 80(662), 94(662), 108(662), 110(921), 114(953), 152, 164, 165
Dietsch, G., 23(165), 132
Dilks, C. H., Jr., 40, 76(321), 138
Dingman, J., 117(968), 166
Dinsmore, S. R., 85(691, 694), 154
Dishman, R. A., 113, 165
Divelbiss, H. N., 30(219), 134
Doali, J. O., 122, 168
Dobbs, R. A., 51, 102(449), 144
Dolansky, V., 68(593), 78(646), 79(655), 150, 152
Dolphin, R. J., 115(954), 165
Domingo, N., 122(1003), 167
Done, J. N., 5(26), 13, 14, 16(85), 23(149), 36, 126, 129, 131
Dowty, B. J., 57(493), 145
Drawert, F., 100(810), 159
Dreiling, R., 57(491), 76(491), 145
Drobishev, V. I., 105(871), 162
Dubsky, H., 49(414), 50, 142, 143
Duch, D. S., 106, 163
Duke, P. D., 53, 144
Dunham, E. W., 96, 158
Dunks, G. B., 102(839), 160
Dupre, G. D., 61(552, 553), 148
Dustin, D. F., 102(838, 839), 160
Duvdevani, I., 59(521), 146

E

Ebisui, S., 102(840), 160
Ecker, E., 23(131, 165, 167), 58(508), 75(634), 131, 132, 146, 151
Ede, P. S., 113, 165
Egan, B. Z., 106, 163
Eisenbeiss, F., 23(150), 109, 131, 164
Ellis, J. P., Jr., 72(610, 611), 150
Ellis, S. R., 57(495), 94(495), 146
Engelhardt, H., 15(90), 17, 18, 39(98, 315), 40(320),

[Engelhardt, H.]
43(335), 64(98), 111(97, 315), 118(97), 121(320), 129, 138, 139
Eon, C., 6(33), 126
Epley, J. A., 85(701), 87(701), 154
Erdahl, W. L., 49(430), 143
Erdel, G., 37(303), 138
Erdody, P., 120(989), 167
Ertingshausen, G., 60(538), 71, 147
Esser, R. J. E., 46(380), 141
Estep, P. E., 78(648), 152
Evans, F. J., 117, 166
Evans, J. E., 93(754), 157
Evans, W. J., 102, 160

F

Fairwell, T., 57(494), 100(494), 146
Fallick, G., 23(132, 151), 131
Feil, V. J., 89(719), 155
Felix, A. M., 72(617, 618), 151
Felton, H., 23(171), 28(171), 30(171), 45(171), 132
Figert, D. M., 98(792), 158
Fishbein, L., 121, 167
Fishman, M. L., 61(558), 148
Fitzpatrick, F. A., 117, 118, 166
Fleischer, J., 30(224), 134
Florsheim, W. H., 74(625), 151
Ford, D. L., 56(489), 145
Forina, M., 53(464), 144
Foti, S. C., 101(822), 160
Fowlis, I. A., 49(409), 142
Frache, R., 101(817), 159
Frank, G., 72(622), 151
Freeman, D. E., 74(625), 151
Freeman, M. L., 85(688, 689), 154
Frei, R. W., 30, 38, 52, 69, 71(605), 72(454), 108(454), 134, 138, 144, 150
Fritz, J. S., 101(819), 102(833, 835, 841), 111(926), 159, 160, 164
Frolov, I., 14(87), 16, 75(87), 116(95), 129
Frost, J. H., 82, 153
Fujimura, K., 7(40-42), 127
Funasaka, W., 7, 127

G

Gabriel, T. F., 105(857), 106, 161, 163
Gallei, E., 37(302, 303), 138
Garcia, J. B., Jr., 72(610, 611), 150
Garden, S. A., 101(830), 160
Gauchel, F. D., 93(755), 105(894), 157, 163
Gauchel, V. G., 93(755), 105(894), 157, 163
Geiss, F., 14(89), 18(104), 129, 130
Gell, J., 105(877), 162
Georgiadis, A. G., 72(619), 151
Gere, D., 102(842, 843), 105(890), 106(900), 161, 163
Gerlach, H. O., 13(83), 18(83, 102), 74(102), 129, 130
Giddings, J. C., 1(5, 9), 9(5, 9), 11(5, 9, 71, 72), 19(107, 108), 21(5), 43(336), 125, 128, 130, 139
Gil-Av, E., 96(781), 100(781), 158
Gilbert, T. W., 51, 102(449), 144
Gilding, 23(158), 132
Gill, J. M., 61, 105(890), 148, 163
Gilpin, R. K., 36(287), 137
Giuffrida, L., 79, 152
Glover, C. A., 80(665), 152
Glumer, M., 50(440), 143
Goley, M. J. E., 21(123), 130
Goldman, P., 85(701), 87(701), 154
Goldstein, G., 59(511), 146
Golkiewicz, W., 7(39), 127
Goostree, B. E., 30(219), 134
Gouw, T. H., 28(197, 207),

[Gouw, T. H.]
30(222), 46(207), 63(568), 77(641), 133, 134, 149, 152
Govan, H. K., 91(735), 156
Grabowski, Z. R., 47(391), 141
Grachev, M. A., 106(904), 163
Grady, L. T., 97(785), 158
Graffeo, A. P., 73, 151
Green, J. G., 105(895, 896), 163
Greenwood, J. M., 101(828, 829), 160
Gregson, K., 61(556), 148
Greve, H., 105(888), 162
Griffin, G. W., 57(493), 145
Grohmann, K., 38(307, 308), 75(307), 138
Grosjean, M., 56(488), 145
Grubisic-Gallot, Z., 4(20), 126
Grunes, D. L., 67(591), 150
Grushka, E., 19(114), 38, 72(309), 130, 138
Gudzinowicz, B. J., 111(930), 164
Guillemin, C. L., 34(275), 136
Guinand, G. G., 98(792), 158
Guiochon, G., 6(33), 9(49, 50), 23(133, 164), 43(339), 63(573), 77(642), 79(654), 126, 127, 131, 132, 139, 149, 152
Gurkin, M., 52(450), 144

H

Haag, A., 73(623), 151
Haahti, E., 49(406), 142
Hadden, N., 31(238), 42(324), 135, 139
Haderka, S., 56, 145
Halász, I., 6(29), 10(53), 12, 13, 15(80, 91), 16(80), 17(98), 18(83, 101, 102), 21(142), 28(208, 209), 29, 31(236), 32(80), 36, 37(300), 39(98), 40(320), 43(349), 46(349), 48, 64(98), 74(102), 121(320), 126-130, 134, 135, 137, 138, 140, 142
Haller, W., 32(250), 135
Hamilton, P. B., 1, 9(52), 13, 46(383-386), 125, 127, 141
Hampai, A., 72, 151
Hana, K., 56, 145
Hanai, T., 7(40-42), 127
Hansen, G., 100(813), 159
Hansen, P. W., 61(565), 148
Hanson, C. V., 103, 161
Hartman, C. H., 44(353), 140
Harven, H., 4(21), 114(21), 126
Harvey, C. L., 106(902), 163
Hashimoto, T., 102(840), 160
Hastings, C. R., 36(292, 293), 137
Hatano, H., 23(134, 135, 168), 52, 63(571, 572), 83(678), 105(893), 131, 132, 144, 149, 153, 163
Hauton, J. C., 122(1001-1003), 167
Hawkes, S. J., 19(105, 109), 31(105), 130
Hawthorne, M. F., 102(837-839), 160
Hayashi, T., 101(824), 160
Hays, S. E., 97(785), 158
Heacock, R. A., 71(605), 150
Heatherly, D. E., 106, 163
Heckles, J. S., 113(943), 165
Heftmann, E., 125
Hegenauer, J., 65(581), 149
Heinekey, D. M., 49(422), 142
Heitz, W., 31, 57(502), 111(933), 135, 146, 164
Hemmingsen, L., 61(565), 148
Henderson, T. R., 87, 155
Hendricks, P. J. H., 8(46), 32(46), 35(46), 123, 127, 168
Hendrickson, J. F., 113(945, 946), 165
Hendrickx, H. K., 90(726), 155
Henry, R. A., 23(172), 29, 57(500), 58(509), 67(588), 80(662), 91(739), 94(662), 99(794), 105(858), 108(509, 662), 110(921), 114, 118(983), 132, 146, 149,

[Henry, R. A.]
152, 156, 158, 161, 164, 165, 167
Henselman, J., 61(550), 148
Herod, J., 105(880), 162
Hesse, C., 115, 166
Hesse, G., 43(335), 139
Hettinger, J. D., 61, 148
Hildebrand, J. H., 4(18), 126
Hill, J. M., 37, 138
Hinsvark, O. N., 90(728, 729), 156
Hirata, Y., 23(136), 131
Hobbs, J. S., 83, 99, 153, 159
Holman, M., 106(902), 163
Holt, R. F., 89(722), 155
Honda, M., 53(463), 144
Hooker, R. P., 28(202), 30(202), 133
Hope, J., 52(457), 144
Horgan, D. F., 10(57), 18(57), 34(274), 41, 74(274), 108(57), 127, 136, 139
Horne, D. S., 9(48), 12(48, 78), 127, 128
Horvath, C. G., 5, 8(43), 11(43), 33, 34, 39, 46(378), 103-105, 126, 127, 136, 141, 161
Horwitz, E. P., 102, 160
Howell, R. R., 85(700), 87, 154
Howery, D. G., 43(348), 140
Hövermann, W., 66, 115, 149, 166
Hsieh, D. P. H., 88(718), 155
Hsu, T. J., 92(746), 156
Hubard, J., 61(552-554), 148
Huber, J. F. K., 4, 5(25), 7(25), 8(44, 45), 11(44), 18(45), 23(137), 32(255), 41, 43(343), 45(343), 48(343), 53(465), 55(475), 58, 96, 101(827), 102(832), 111, 113, 114, 126, 127, 131, 136, 140, 144-146, 158, 160, 165
Hudson, B. G., 60(539), 147
Hudson, D. R., 58(509), 65(579), 74(579), 97(579), 108(509), 146, 149
Hudson, J. S., 82(673), 153
Hughes, D. W., 96(778), 158
Hulsman, J. A. R. J., 4(19), 8(45), 18(45), 41(19), 114(949-951), 114(19), 126, 127, 165
Hums, N., 100(812), 159
Hunt, J. A., 54, 145
Hupe, K. P., 23(173), 43(351), 50, 77(173), 132, 140, 143

I

Imai, K., 72(615), 151
Inaba, T., 93(753), 157
Inczedy, J., 23(144, 145), 43(330), 131, 139
Inglis, A. S., 71, 150
Irreverre, F., 88(717), 155
Ishiguro, S., 60(530), 147
Ives, N. F., 79, 152
Izuhara, M., 102(840), 160

J

Jackson, A., 55(477), 145
Jacobsen, A. M., 89(719), 155
Jainchill, J. L., 85(689), 154
James, A. T., 1(7), 49(407), 55(476), 125, 142, 145
Janák, J., 60(531), 78(649), 147, 152
Jandera, P., 66(583), 149
Jane, I., 94(757), 157
Jansen, J. M., 19(118), 61(118), 85(118), 130
Jardy, A., 10(61), 13, 18(61, 81), 128, 129
Jefferson, R. T., 38, 138
Jennings, E. C., 118(976), 166
Jentoft, R. E., 28(197, 207), 30(222), 46(207), 63(568), 77(641), 133, 134, 149, 152
Johnson, D. C., 56, 102(480), 145
Johnson, H. W., Jr., 48, 49, 141
Johnson, J. F., 31(234), 32(247, 253), 113(942), 135, 136, 165

Johnson, P. E., 26(187), 60(187), 133
Johnson, W. F., 27(190), 46(190), 60(190), 85(190, 695), 133, 154
Johnson, W. T., 30(221), 85(221), 134
Jolley, R. L., 19(115), 27(190), 46(190), 60(190), 61(118), 85, 118(985), 119(985), 130, 133, 154, 167
Jolliffee, G. H., 71(604), 150
Jones, C. J., 102(838), 160
Jones, G., Jr., 76(638), 85(692, 695, 696), 152, 154
Jones, R. K., 87, 155
Joynes, P. L., 53, 144
Juhasz, A. A., 122, 168
Junowicz, E., 105(889), 163
Jurand, J., 70(602, 603), 150

K

Kahn, H. L., 23(138), 99(138), 131
Kaiser, K. J., 65, 149
Kalab, P., 56, 145
Kallet, E. A., 52(450), 144
Karasek, F. W., 23(139), 30(229), 36(285), 131, 135, 137
Karger, B. L., 9(47), 17(98), 19(112), 28, 29, 30(215), 31, 39, 40(320), 61(567), 64(98), 73(623), 77, 111(315), 118, 121(320, 996), 127, 129, 130, 134, 138, 149, 151, 152, 166, 167
Karmen, A., 49(408, 416, 426), 142, 143
Karol, P. J., 101(820), 159
Karr, C., Jr., 1, 78, 125, 152
Katz, S., 27, 46(388), 68, 85(682, 690-692, 699), 122, 133, 141, 150, 153, 154, 168
Kawamura, J., 23(140), 131
Kawazu, K., 102(841), 160
Kelemen, S. P., 46(381), 105, 141
Keller, H. E., 61(567), 149
Keller, R. A., 6(30), 43(336, 337), 126, 139
Kelley, W. N., 87, 155
Kelmers, A. D., 106, 163
Kennard, W., 56(489), 145
Kennedy, G. J., 5(26), 34, 35, 126, 136
Kennedy, W. P., 105(849), 161
Kern, H., 117(966), 166
Kersten, T. E., 49(427), 143
Kesler, R. B., 82, 153
Ketelle, B. H., 101(823), 160
Khopina, V. V., 7(37), 127
Khym, J. X., 118(985), 119(985), 167
King, R. H., 97(785), 158
King, R. N., 27(192), 28(198), 30(198), 42(192), 75(198), 77(198), 108(198), 111(939), 133, 165
Kirkland, J. J., 10(59, 60), 11(59), 13, 14, 15(59), 16(59), 17, 23(152), 30(230), 31(88), 32(60, 88, 254), 33, 34(60, 230, 254), 35(60, 277), 36(254, 286, 294), 40, 43(338), 46(382), 64, 65(230), 68(260), 69, 76(254, 321), 82(668), 89, 95(767), 105(853), 106(88), 107, 108(915), 113(60), 121(88), 128, 129, 131, 135-139, 141, 153, 155, 157, 161, 164
Kiselev, A. V., 7(36, 37), 14(36), 75, 127, 129
Kissinger, P. T., 57(491), 76, 145
Klatyk, K., 111(933), 164
Kleber, W., 100(812), 159
Klein, H. R., 97(785), 158
Klein, P., 14, 129
Klimisch, H. J., 78, 79(652), 152
Klose, A., 14(89), 129
Kluckhohn, J. H., 94(758), 157
Knox, J. H., 1(8, 10), 5(23, 24, 26), 9(48), 10(54),

[Knox, J. H.]
11(65), 12(48, 78), 13-16, 19, 23(149, 159), 34-36, 70(602, 603), 125-132, 136 137, 150
Kochen, W., 23(166), 132
Koen, J. G., 53, 144
Koenders, E. B., 55(474), 145
Koh, C. K., 60, 103(846), 106(545), 148, 161
Kolder, F. F. M., 113(947), 165
Komarova, N. I., 106(904), 163
Koreeda, M., 59(528), 147
Koss, R. F., 88(715), 155
Koszewski, J., 47(391), 141
Koutsky, J. A., 31(235), 135
Kraak, J. C., 101(827), 160
Kraffczyk, F., 111(933), 164
Kram, T. C., 94, 117, 157, 166
Kramer, U., 111(928), 164
Krasny, H. C., 96(775), 157
Kraus, P., 90(729), 156
Krebs, K. F., 35(280), 137
Krejci, M., 30, 43(340), 49(414), 60(226), 76(226), 134, 139, 142
Krol, G. J., 118(984), 167
Kroll, M. G. F., 8(46), 32(46), 35(46), 127
Kroneisen, A., 13(83), 18(83, 102), 43(349), 46(349), 74(102), 129, 130, 140
Kruger, P. M., 49(427), 143
Krzeminski, L. F., 90(724), 155
Kucera, P., 26, 31(241), 105(883), 133, 135, 162
Kulaev, I. S., 105(871), 162
Kuras, M., 79(655), 152
Kuzmin, S. V., 106(904), 163

L

Ladd, F. C., 105(896), 163
Lafont, H., 122(1003), 167
Lairon, D., 122(1002-1004), 167
Lampert, K., 43(333), 139
Lamprecht, W., 67(590), 150
Landgraf, L. M., 61(558), 148
Landgraf, W. C., 118(976), 166
Langille, K. R., 71(605), 150
Lapidus, B. M., 49(426), 143
Larmann, J. P., 65(579), 74(579), 97(579), 149
Larochelle, J., 56, 102(480), 145
Larson, P., 92(746), 156
Laseter, J. L., 57(493), 145
Laskowski, M., 106(906), 163
Latham, D. R., 80, 152
Lattore, J. P., 90, 92, 156
Laub, E., 98, 158
Lawrence, J. F., 52(452-454, 457), 72(454), 108(454), 144
Lawrence, J. G., 12(75), 25, 43(179), 49, 56(490), 83(679), 99, 122, 128, 132, 142, 145, 153, 159, 167
Ledford, C. J., 80, 153
Lee, J. C., 105(849), 161
Lee, N. E., 32, 85(684, 705), 136, 153, 154
Lee, Y. C., 83, 153
Le Febre, H. A., 48(400), 141
Leicht, R. E., 23(152), 33(262), 34, 42(325), 108(325), 131, 136, 139
Leimgruber, W., 52(455), 72(613, 614, 455), 144, 151
Le Page, M., 34(275), 111(936), 136, 165
Leslie, R. C., 23(160), 132
Levangie, R. F., 28(198), 30(198), 59(525, 526), 75(198), 78(198), 96, 108(198), 111(525, 938), 112(940), 113(940), 133, 146, 158, 165
Liao, T. H., 72, 151
Liebnberg, C. J., 43(331), 101(331), 139
Limpert, R. J., 34(272), 136
Lindley, H., 30, 60(228), 134
Lipsky, S. R., 5, 8(43), 11(43), 33(264), 46(378), 104(264), 105, 126, 127, 136, 141
List, D. M., 98, 158
Little, J. N., 10(57), 18(57),

[Little, J. N.]
32(243, 246), 34, 41, 74(274), 108(57), 111(243, 931, 932, 934), 127, 135, 136, 139, 164
Lochmüller, C. H., 28(202), 30(202), 133
Locke, D. C., 2, 5(22, 27), 6, 36, 39, 77, 126, 137, 138
Lodge, B. A., 115(956), 166
Loffey, J. W., 72(619), 151
Loheac, J., 9(50), 77(642), 79, 127, 152
Longbottom, J. E., 67(589), 150
Lovins, R. E., 57, 94(495), 100(494), 146
Lötscher, K., 117(966), 166
Luckhurst, G. R., 6, 126

M

Ma, T. S., 57(505), 146
MacDonald, A., 53, 144
MacDonald, F. R., 19(111), 20, 29(211), 30(216), 105(854, 890), 130, 134, 161, 163
Machin, A. F., 28(205), 110, 133, 164
Mac Neil, J. D., 71(605), 150
Mader, W. J., 97(785), 158
Maggs, R. J., 25(180, 181), 43(181), 49(409-411, 423), 53, 132, 142, 144
Majors, R. E., 13-15, 16(84), 19(111), 20, 32(258), 33(261), 34(258), 36, 40(319), 75, 108(319), 110(631), 117(967), 121(631), 129, 130, 136, 138, 151, 166
Maley, L. E., 47(392), 141
Mansurova, S. E., 105(871), 162
Maresca, R. A., 88(714), 155
Marinello, J. M., 90(729), 156
Marsden, N. V. B., 61, 148
Marsh, A. C., 99, 158
Martin, A. J. P., 1(6, 7), 43(347), 55(476), 125, 140, 145
Martin, G. E., 98, 158
Martin, M., 9(49, 50), 23(164), 63(573), 77(642), 79(654), 127, 132, 149, 152
Martin, W. E., 88(716), 155
Martinez, B., 105(887), 162
Martinsson, E., 82(675), 153
Martinu, V., 78(649), 152
Martire, D. E., 2, 6, 126
Matsumoto, T., 7(40), 127
Mauger, A. B., 96, 158
Maurer, T., 59(521), 146
Maxwell, W. B., 120(989), 167
Mayland, H. F., 67(591), 150
McCready, R. M., 82(673), 153
McGilveray, I. J., 97(783), 158
McGuiness, E. T., 54, 145
McKay, J. F., 80, 153
McKinney, C., 57(494, 495), 94(495), 100(494), 146
McLaren, L., 9(48), 12(48), 12(78), 43(336), 127, 128, 139
McNair, H. M., 23(157), 38, 64(312), 66(312), 132, 138
Meijers, C. A. M., 4(19), 41(19), 113(949, 951, 952), 126, 165
Menzorova, N. I., 106(904), 163
Merle, W., 100(810), 159
Michalewsky, J. E., 105(857), 106(901-903), 161, 163
Miech, R. P., 91(733), 105(881, 882), 156, 162
Mikes, F., 96(781), 100(781), 158
Miles, R., 60(541), 147
Miller, J. M., 113(947), 165
Mironova, I. V., 16(95), 116(95), 129
Mochida, E., 52(461), 144
Modl, V. A., 10(62), 18(62), 101(826), 128, 160
Moffitt, E. A., 85(697), 154
Mollica, J. A., 91(735), 96, 115(961, 962), 118, 156, 157, 165, 166
Molyneux, R. J., 100(809), 159
Montet, J. C., 122(1001), 167
Moore, J. C., 32(244),

[Moore, J. C.]
113(946), 135, 165
Moore, S., 46(377), 141
Mopper, K., 82(671), 153
Moretti, G., 115(960-962), 166
Morie, G. P., 80(665), 153
Moriizumi, S., 60(530), 147
Morris, B. M., 10(62), 18(62), 128
Morris, C. R., 28(205), 133
Morrow, G., 88(714), 155
Mostecky, J., 68(593), 78(646), 79(655), 150, 152
Mrochek, J., 61(560), 85(685), 686, 693, 694), 87, 148, 153, 154
Mulder, J. L., 57(503), 146
Munk, M. N., 30(216), 43(345), 45(364), 50(445), 51, 105(854), 134, 140, 143, 161
Munro, H. N., 105(876), 162
Murakami, F., 83(678), 105(893), 153, 163
Murgia, E., 92, 156
Murphy, F. J., 110(921), 164
Muto, G., 56, 145
Muusze, R. G., 96, 158
Myers, N. N., 43(336), 139

N

Naefe, M., 12, 13, 15(77, 80, 91), 16(80), 32(80), 128, 129
Nakamura, S., 60, 147
Nakanishi, K., 59(528), 147
Nalbone, G., 122(1003), 167
Naono, T., 50(441), 143
Necasova, M., 36(281), 137
Neff, A. W., 90(724), 155
Nelson, D. J., 96(775), 157
Nelson, J. A., 96(774), 105(774), 157
Nelson, J. F., 97(789), 158
Nibbelink, D. W., 90(725), 155
Nicholls, P. W., 71, 150
Nickless, G., 36(284), 137
Nikarri, T., 49(406), 142
Nilsson, O., 47, 141
Nocho, R. E., 88(714), 155
Nota, G., 49(413), 142
Novosel, B., 6(33), 126
Novotny, M., 38, 75(307), 138
Nunes, I. M., 91(735), 156
Nunley, C. E., 105(895), 163
Nussbaum, A. L., 106(902), 163
Nyamah, D., 118(979), 166

O

Oakes, P. L., 49(428), 143
Obermeyer, B. D., 61(563), 148
Odell, V., 60(539), 147
Oesterhelt, G., 120(991), 167
Ohashi, S., 57(492), 145
Ohrt, D. W., 85(694), 154
Oreans, M., 58(508), 146
Oster, H., 43(346), 75(634), 140, 151
Osterdout, H. W., 113(943), 165
Otocka, E. P., 23(141), 36(283), 131, 137
Otterg, E., 82(672), 153

Ö

Öberg, P. A., 61(596), 148
Ötvös, L., 23(125), 63(125), 130

P

Padmanabhan, G. R., 96(771), 157
Pajurek, J., 49(414), 142
Palamand, S. R., 100, 159
Palmer, J. K., 98, 158
Palombari, R., 49(413), 142
Papa, L. J., 65, 149
Parcher, J. F., 10(54), 15, 16, 127
Parks, J. E., Jr., 96(773, 774), 105(773, 774), 105(874, 877-880), 106(899), 157, 162, 163
Parr, W., 38(307), 75(307), 138
Parris, N. A., 58(509), 108(509), 146

Patterson, R. P., 67(591), 150
Pattison, M. H., 49(417), 142
Paulson, G. D., 89(719), 155
Pauplis, W. J., 32(246), 34(246), 111(931, 932, 934), 135, 164
Pease, H. L., 89(722), 155
Pecsok, R. L., 56, 145
Peczon, B., 60(539), 147
Pellizzari, E. D., 52, 144
Pennington, S. N., 105(862), 161
Perry, S. G., 12(79), 32(79), 36(288), 63(569), 75(635), 78(79), 129, 137, 149, 151
Persinger, H. E., 63(574), 65(574), 149
Persson, A., 61(546), 148
Pesek, J. J., 82, 153
Pfadenhauer, E. H., 106(898), 163
Pfaff, K. J., 85(697), 154
Pfitzner, K., 111(933), 164
Phillip, L. J., 52(451), 144
Pickett, J. A., 105(867), 162
Pidacks, C., 34(271), 136
Piel, E. W., 1, 14(15), 28(15), 126
Pitt, W. W., Jr., 19(118), 27(189, 190), 46(190, 387, 388), 60(189, 190), 61(118, 562), 68, 76(638), 85(118, 190, 562, 682, 683, 691, 692, 695, 696), 122(1005), 130, 133, 141, 148, 149, 152-154, 167
Poet, R. B., 94, 157
Polak, H. L., 54, 72, 145
Polesuk, J., 43(348), 140
Polgar, A. G., 49(432), 143
Politzer, I. R., 57(493), 145
Pollard, F. H., 36(284), 137
Popl, M., 68(593), 78(646), 79, 150, 152
Poppe, H., 4(21), 53(465), 114(21), 126, 144
Popplewell, J. A., 82(672), 153
Porcaro, P. J., 110(920), 164
Porter, R. S., 31(234), 135
Pospisilova, N., 43(340), 139
Potthast, H., 23(170), 132
Pound, N. J., 29(213), 96(778), 97(783), 784), 115(956), 134, 158, 166
Pracharova, M., 36(282), 137
Preiss, B., 33(264), 104(264), 136
Pretorius, V., 1(11, 12), 11(69), 19(11, 69, 106), 27(191), 30(225), 49, 50(438), 51(448), 125, 128, 130, 133, 134, 143
Price, C. E., 89(721), 155
Price, L. W., 23(161), 132
Prior, R. L., 67, 150
Privett, O. S., 49(430), 122, 143, 167
Pu, H. H., 94, 157
Pungor, E., 53(467), 144
Purnell, J., 19
Pye, M. J., 49(424), 142

Q

Quano, A., 43(332), 139
Quick, M. P., 28(205), 133
Quillet, R., 56(486), 145
Quinn, E. J., 113(943), 165

R

Rabel, F. M., 38, 75(311), 94(311), 97(311), 138
Rainey, W. T., Jr., 85(693, 707), 154, 155
Rajcsanyi, P. M., 15, 23(125), 30(217), 63(125), 71, 95, 102(844), 129, 130, 134, 150, 157, 161
Rajcsanyi-Kriskovics, E., 102(844), 161
Ramus, S. E., 100(813), 159
Randall, S. S., 55(476), 145
Randau, D., 17, 75(630), 77(96), 110(630), 111(933), 129, 151, 164
Rao, G. H. R., 57(496, 497), 89(497), 146

Rapp, A., 66(585), 149
Raval, D. N., 51, 143
Ravenhill, J. R., 49(407), 142
Ray, S., 38, 69, 138
Refshauge, C., 57(491), 76(491), 145
Reichler, A. S., 60(538), 71(538), 147
Reissenweber, H., 99(797), 159
Richards, P., 92(747), 156
Richter, P. W., 50(438), 143
Rickett, F. E., 57(504), 146
Rijnders, G. W. A., 11(70), 28(206), 128, 134
Ringe, P., 35(276), 137
Ritchie, C. L., 30(228), 60(228), 134
Riumallo, J. A., 105(876), 162
Riva, M., 60(542), 147
Robinson, G. W., 72(612), 151
Robinson, T., 70(601), 150
Rogers, L. B., 28(202, 203), 30(202), 133
Rokushika, S., 83(678), 105(893), 153, 163
Roos, R. W., 93, 157
Rosevear, J. W., 85(697), 154
Ross, J. H., 32(248), 38, 135, 138
Rosset, R., 10(61), 13, 18(61, 81), 128, 129
Roth, M., 52(456), 72, 144, 151
Rouser, G., 122, 167
Rössler, G., 29(214), 134
Rzeszotarski, W., 96, 158

S

Saito, M., 52(461), 144
Saleem, M., 11(65), 19, 128, 130
Salnikow, J., 72(612), 151
Saltman, P., 65(581), 149
Samuelson, O., 82(670, 675), 153
Sass, S., 71(607), 150
Saunders, D. L., 25, 42(177), 56, 78, 132, 145
Scarano, E., 53(464), 144
Schäfer, G., 67(590), 105(868), 150, 162
Schick-Kalb, J., 35(276, 280), 137
Schier, G., 37(305), 138
Schill, G., 23(143), 131
Schlatter, J. E., 49(428), 143
Schlimme, E., 105(868, 869), 162
Schlitt, H., 14(89), 18(103, 104), 129, 130
Schmermund, J. T., 36(295), 137
Schmit, J. A., 26(187), 60(187), 67(588), 69(594), 80, 91(739), 94(662), 99(794), 105(858), 106(908), 108(662, 908, 913), 110(921), 114(953), 118(980, 983), 133, 149, 150, 152, 156, 158, 161, 163-165, 167
Schnabel, C., 115(959), 166
Schneider, J., 83, 153
Schneider, W., 29(214), 134
Schnell, W., 17, 77(96), 129
Scholar, E. M., 96(773), 105(773, 879), 157, 162
Schram, E., 54, 144
Schreiber, J., 59(514), 146
Schrenker, H., 23(169, 173), 77(173), 132
Schulten, H. R., 57(498), 115, 146
Schuppe, E., 23(170), 132
Schurig, V., 96(781), 100(781), 158
Schutte, L., 55(474), 145
Scott, C. D., 19(115, 116, 118), 27, 30(221), 31(240), 32, 33(240), 35, 43(329), 46(190, 387, 388), 59(511), 60(189, 190), 61(115, 116, 118, 561, 562), 84, 85, 91, 118(985), 119(985), 122(1005), 130, 133-137, 139, 141, 146, 148, 153-156, 167, 168
Scott, C. G., 95(764), 105(883), 157, 162
Scott, R. L., 4(18), 126

Scott, R. P. W., 12(73, 75), 21, 25, 26, 31(232, 241), 38, 43(179), 49, 50(435, 437), 56(490), 72(309), 105(883), 128, 130, 132, 133, 135, 138, 142, 143, 145, 162
Sears, R. W., 29(213), 97(783, 784), 134, 158
Sebestian, I., 36(291), 37(300), 137, 138
Segal, L., 125
Segura, R., 50(442), 143
Seiber, J. N., 88(718), 155
Seibert, E. E., 49(431, 432), 143
Senft, A. W., 105(874, 881), 162
Senft, D., 105(881), 162
Seyler, R., 101(830), 160
Seymour, M. D., 101(819), 102(833-835), 159, 160
Shanks, W. H., 101(828), 160
Shargel, L., 88, 155
Shellard, E. J., 71(604), 150
Shelman, D. L., 94(759), 157
Shepherd, J. A., 90(725), 155
Shimizu, T., 105(861), 161
Shmukler, H. W., 42(326), 105(850-852), 132, 161
Shu, C. K., 99(793), 158
Shubiak, P., 110(920), 164
Sickafoose, J. P., 101(819), 159
Sie, S. T., 11(70), 15, 28(206), 32(93), 39(93), 40(93), 74(93), 128, 129, 134
Siebert, K. J., 101, 159
Sieper, H., 109, 164
Sieswerda, G. B., 54, 58(507), 72, 145, 146
Siggia, S., 69(595), 70(601), 113, 117, 150, 165, 166
Simkin, A., 89(720), 155
Simpson, D., 23(142), 131
Simpson, D. W., 10(658), 128
Siniscalchi, P., 115(961), 166
Sirec, M., 30(226), 60(226), 76(226), 134
Sisson, D. H., 10(62), 18(62), 101, 128, 160
Sivertson, J. N., 61(557), 148
Skelly, N. E., 67(587), 97, 149, 158
Sleight, R. B., 74, 80, 110, 151, 152
Smith, F. W., 67(591), 150
Smith J. B., 91(735), 156
Smuts, T. W., 1(11, 12), 11(68, 69), 19(11, 69, 106), 27(191), 30, 50(438), 51, 125, 128, 130, 133, 134, 143
Snyder, L. R., 7, 10, 11(66, 67), 12-14, 17, 18(100), 19(82, 113), 21, 24(175, 176), 25, 28(196), 32, 42, 43(55, 175, 176, 337), 68, 78, 126-130, 132, 133, 139
Soczewinsky, E., 7(39), 127
Solms, D. J., 27, 30(225), 133, 134
Spackmann, D. H., 46(377), 141
Sparacino, C. M., 52, 144
Spencer, J. H., 105(889), 163
Spitz, H. D., 61(557), 148
Stafford, D. T., 50(436), 143
Stahl, E., 98, 158
Stahl, K. W., 23(170), 67, 105(868, 869), 132, 150, 162
Stasink, L., 106(906), 163
Stegink, L. D., 90(725), 155
Stegman, R. J., 105(874), 162
Stehl, R. H., 36(290), 82(669), 137, 153
Stein, S., 52(455), 72(455, 615, 616), 144, 151
Stein, W. H., 46(377), 141
Stenberg, V. I., 48, 142
Stermitz, F. R., 91, 156
Sternberg, J. C., 56(487), 145
Stevens, R. H., 49(418), 142
Stevenson, R. L., 77, 87, 92(740), 100, 119(986), 152, 155, 156, 159, 167
Stewart, H. N. M., 12(79), 32(79), 36(288), 78, 128, 137
Stewart, I., 99(795), 158
Stillman, R., 57(505), 146
Stolyhwo, A., 49, 122, 143, 167
Stone, J., 72(615, 615), 151
Stouffer, J. E., 49, 143

Straube, B., 35(276), 137
Strelow, F. W. E., 43(331), 101(331, 818), 139, 159
Strohl, J. H., 53(462), 144
Stross, F. H., 49(431, 432), 143
Strubert, W., 32(259), 72(622), 78, 136, 151, 152
Strusz, R., 96(771), 118, 157, 166
Stutz, M. H., 71(607), 150
Sutherland, W. J. A., 30(228), 60(228), 134
Sweeney, J. P., 99, 158
Sybilska, D., 47(391), 141
Synge, R. L. M., 1(6), 125
Szepesvary, E., 53(467), 144

T

Tabor, C. W., 88(717), 155
Tabor, H., 88(717), 155
Takagi, T., 102(840), 160
Takata, Y., 56, 145
Takemori, Y., 53(463), 144
Talley, C. P., 69, 150
Tan, M., 59(521), 146
Taylor, L., 82(672), 153
Telepchak, M. J., 32(251), 79, 136, 152
Tengi, J., 72(613), 151
Terkelsen, G., 72(617, 618), 151
Tesarik, K., 36, 56, 145
Thacker, L. H., 46, 52, 85(699), 141, 144, 154
Thieband, M., 69(598), 150
Thomas, R. D., 91, 156
Thomas, W., 37(301), 138
Thompson, J. N., 120(989), 167
Thorn, W., 105(888), 162
Thornton, J. I., 94(759, 760), 157
Tolbert, G. D., 57(495), 94(495), 146
Toporak, M., 52(451), 144
Toshida, K., 23(126), 130
Touchstone, J. C., 115, 166
Trefz, F., 23(166), 132
Tung, M. C., 105(882), 162
Turner, L. P., 65, 149

U

Uden, P. C., 36(284), 137
Udenfriend, S., 52(455), 72(455, 615, 616), 144, 151
Ullner, J., 57(502), 146
Unger, K., 35, 37, 118(979), 137, 138, 166
Uziel, M., 61, 103, 106(545), 148, 161

V

Van Damme, S., 75(634), 151
Van den Hoed, N., 15(93), 32(93), 39(93), 40(93), 43(334), 74(93), 129, 138 139
Vanderkerckhove, P., 90(726), 155
Van der Linden, R., 58(508), 146
Van de Weerdhof, K., 99, 159
Van Dijk, J. H., 49(425), 142
Vanheertum, R., 23(146), 100, 131, 159
Van Kreveld, M. E., 43(334), 139
Van Niekerk, F. A., 1(12), 30(225), 51(448), 125, 134 143
Van Niekerk, P. J., 120(990), 167
Van Rensberg, J. F. J., 49, 143
Van Urk-Schoen, A. M., 55(475), 58(507), 102(832), 145, 146, 160
Vasvari, G., 35, 137
Vaughan, C. G., 80(663), 152
Vavich, J. M., 85(700), 87, 154
Vecchi, M., 120, 167
Vedso, S., 61, 564), 148
Veening, H., 23(162, 163), 43(341, 342), 47(341), 76, 101, 132, 139, 152, 160
Versino, B., 18(103), 130

Verzele, M., 100(811), 159
Vesely, J., 120(991), 167
Vespalec, R., 30(226), 56, 60(226), 76(226), 134, 145
Vestergaard, P., 61, 148
Vigh, Gy., 23(144, 145), 131
Virkola, P., 105(870), 161
Visek, J., 67(591), 150
Vivilecchia, R., 69(598), 150
Vlasov, V. V., 106, 163
Von S. Toerin, F., 43(331), 101(331), 139
Vorobyeva, R. G., 16(95), 116(95), 129
Vratny, P., 90(727), 155

W

Walker, V. E., 30(221), 85(221), 134
Walkling, P., 10(53), 12(74, 76), 13, 18(83, 102), 31(236), 48, 74(102), 127-130, 135
Walradt, J. P., 99(793), 158
Walton, H. F., 92(746, 747), 156
Warner, W. C., 1(14), 78(648), 125, 152
Warren, K. S., 85(689, 698, 706), 154
Waskiewicz, R. D., 70(601), 150
Watanabe, S., 23(134), 52(461), 83(678), 131, 144, 153
Waters, J. L., 10, 18(57), 32(246), 34(246), 59(522-524), 108(57), 111(931, 932, 934), 127, 135, 146, 164
Watson, E. S., 45(376), 47(393), 141
Weber, D. J., 95, 157
Wegener, L., 60(539), 147
Weigand, N., 17(97), 111(97), 118(97), 129
Weigele, M., 52(455), 72(613, 614), 72(455), 144, 151
Weiss, G., 59(528), 147
Wheals, B. B., 80(663), 94(757), 152, 157
Wheaton, R. M., 10(58), 128
Wheaton, T. A., 99(795), 158
Whitehouse, M. J., 80(663), 152
Whitlock, L. R., 31(234), 135
Wiedemann, H., 15(90), 18, 129
Wiersma, R. J., 102(838), 160
Wiersum, M. L., 99(797), 159
Wildanger, W. A., 97(788), 158
Wilkins, T., 12(73), 31(232), 128, 135
Willeford, B. R., 101(828-830), 160
Williams, A. D., 74(625), 151
Williams, R. C., 58(509), 65(579), 74(579), 80(662), 94(662), 97(579), 99(794), 105(858), 108(509, 662), 118(980, 983), 146, 149, 152, 158, 161, 167
Willis, R. B., 111(926), 164
Wish, L., 101(822), 160
Wittwer, J. D., Jr., 94(758), 157
Wohlleben, G., 32, 135
Wolen, R. L., 61(563), 148
Wolf, J. P., III, 58, 146
Wong, Y., 100(809), 159
Woodward, R. B., 59(512), 146
Wortmann, W., 115(958, 959), 166
Wu, C. M., 82(673), 153
Wu, C. Y., 69(595), 70, 150
Wyatt, D. K., 97(785), 158
Wyngaarden, J. B., 87(708), 155

Y

Yamabe, T., 23(147), 101(824), 131, 160
Yamada, T., 60(530), 147
Yamamoto, M., 102(840), 160
Yamamoto, Y., 52(461), 102(840), 160
Yashin, Y. I., 14(87), 16(95), 75(87), 116(95), 129
Yost, R. W., 23(153), 131
Young, D. S., 85(701-703), 87(701-703), 154

Young, T. E., 25(180), 49(410), 132, 142
Yoza, N., 57(492), 145

Z

Zahn, R. K., 105, 162, 163
Zaylskie, R. G., 89(719), 155
Zazulak, W., 90(728, 729), 156
Zbrozek, J., 90(727), 155
Zenie, F. H., 23(151), 131
Ziegler, A., 66(585), 149
Ziegler, D. C., 113(943), 165
Zimmere, R. O., 97(785), 158
Zimmerman, T. P., 96(775), 157
Zlatkis, A., 50(442), 143

SUBJECT INDEX

A

Abate, 110
Abscisic acid, 59
Acenaphtalene, 78
Acetaminophen, 92
Acetanilide, 92
Acetic acid, 87
Acetophenone, 34, 64
5-Acetylamino-6-amino-3-
methyluracil, 87
Acids, in biological fluids, 85
Actinides, 101
Actinomycins, 96
Adenine, 87
Adsorbent:
  porosity, effect on column
  performance in LSC, 14
  water content of, 14
Adsorption mechanism, of aromatics on anion exchange
resin, 7
Aflatoxins:
  from peanut butter extract,
  97
  reverse-phase HSLC of, 88, 89
Agmatine, 88
Alcohols:
  aliphatic, 63
  n-amyl, 64
[Alcohols]
  aromatic, 64
  benzyl, 31, 64
  n-butyl, 64
  tert-butyl, 64
  cinnamyl, 64, 65
  α-α'-dimethyl benzyl, 65
  ethyl, 64
  isopropyl, 64, 76
  methyl, 63
  α-methyl benzyl, 64
  2-phenylethyl, 64
Aldehydes, aliphatic, as 2,4-
dinitrophenylhydrazones,
65
Aldomet, 89
Aldosterone, 115
Alkali metals, traces of, 101
Alkaline ferricyanide method,
83
Alkaloids, 68-71
  adsorption energies of, 68
  chincona, 70
  ergot, 71
  opium, 70
  oxindole, 71
  purine, 69
  tropane, 71
  Veratrum Viride, 68
Alkylbenzenes, 79

[Alkylbenzenes]
  chromatographic behaviour on Permaphase ODS, 80
Alkyl parabens, 97
Alkyl phthalates, 121
Allulose, 85
Alumina, water deactivated, 79
Amines:
  aliphatic, in urine, 88
  aromatic, 65
Aminex, 65, 89
Amino acids:
  analysis, 34, 60, 61
  analyzers, fully automated, 71
  in biological fluids, 85
  in blood plasma and serum, 90
  dansyl derivatives of, 72
  ^{14}C-labeled, 72
  phenylthiohydantoine derivatives of, 72, 73
  single column analysis of, 72
α-Amino acids, detection of, 72
Aminoacyl-tRNA, tracer labeled, 106
2-Aminobenzimidazole, 89
α-Amino-n-butyric acid, 71, 72
ε-Aminocaproic acid, in body fluids, 90
Amino metabolism, study of, 90
Amino sugars, 72
Amniotic fluid, analysis of, 85
Analgesics, 91, 92
  effect of cross linking and counterions on separation of, 92
  influence of the ionic strength of the mobile phase on separation of, 91
Analysis, time of, 8, 9
Androgenes, 114
Anilines, 74, 76
  mechanisms of separation of, 74
Anisaldehyde, 99
Antazoline, 96
Anthracene, 37, 77, 78, 80
Anthraquinones, 80, 81
  effect of temperature on separation of, 81
Anthraxanthin, 99
Antioxidants, 75, 110
  phenolic, 110
Antitussives, 92
Arabinose, 82, 83, 85
Aromatic hydrocarbons, 77-80
  alcohol content of the mobile phase in the separation of, 79
  from pitch oils, 78
  polycyclic, 77, 79
  polynuclear, 61
    chromatographic behavior on Permaphase ODS, 80
Aromatic sulfonates, 82
Arylglycosides, anomeric, 83
Ascorbic acid, 65, 97, 119
Aspirin, 91, 92
Asthma tablet, analysis of, 92
Atropine, 71
Auto exhaust condensate, analysis of, 80
Automated HSIEC of carbohydrates, 82
Automatic:
  data acquisition, 61
  phosphorus analysis, 122
  sample loader, 30
Automation:
  levels of, in HSLC, 60
  in routine analyses, 60
Azaheterocyclics, 38
Azobenzene, 75, 76

B

Barbiturates, 92-94
  metabolites of, 92
Benomyl, residues of, 89
3,4-Benzacridine, 25
Benzaldehyde, 65
Benzanilide, 64
1,2-Benzanthracene, 77-79
Benzene, 25, 35, 37, 64, 77, 80

[Benzene]
  nitro derivatives of, partition coefficients of, 4
  sulfonic acids, 68
Benzenes, partition coefficients of alkyl, 4
Benzimidazoles, substituted, 69
Benzodiazepines, 95
Benzo(b)fluoranthene, 79
Benzoic acid, 66, 67
Benzpyrene, isomers, 53, 77-80
Biobeads, 113
Bio Gel, 83, 89
Biogenic amines, 76
Biphenyl, 97
Blood serum, analysis of, 85
Body-fluid mixtures, analysis of, 61
Bondapak, 80, 94, 99, 117
BOP, 16, 17, 31, 34, 64, 65, 74, 75, 77, 108, 111, 114-117, 120, 121
Bromobenzenes, 80
m-Bromomethylbenzoate, 78
Brucine, 69
Butyric acid, 87

C

Cadaverine, 88
Cadmium(II), 102
Caffeine, 69, 70, 91, 92
Capacity factor, 7
  dependence on solvent strength, 5
  optimal, 5
Carbaryl, 108
Carbohydrates, 80-84
  in biological fluids, 84, 85
  in milk, 99
  relationship between temperature and resolution of, 82
Carbonyl compounds, in auto exhaust, 65
Carbowax, 69, 77
Carboxylic acids, 67
  aliphatic, 65
[Carboxylic acids]
  amino, 71
  benzene-poly-, 66
  nitrosubstituted aromatic, 65
Carotene, 99
  stereoisomers, 99
Carotenoid, 99
Cephaeline, 68
Cerate oxidation of fluorimetry, 68
Cerebrospinal fluid, analyis of, 85
Chlorinated biphenyls, 26
2-Chloroaniline, 76
3-Chloroaniline, 76
Chlorobenzene, 25, 34, 80
Chlorpromazine, 96
Cholesterol, 122
Cholesteryl phenylacetate, 59
Chrysene, 79
Cinchonine, 70
Cinnamylic acid, 66
Citric acid, 67
Citrulline, 71, 72
Clinical analysis, automated methods for, 61
Clopidol, 97
Coating:
  of prepacked columns, 39
  solvent evaporation method of, 39
  steady-state method of, 40
Cocaine, 70, 95
Codeine, 70, 95
Column:
  automated regeneration of, 27
  calculation of minimum value of infinite diameter, 16
  cellulose acetate, 79
  coated in situ, 65
  coiling, influence on efficiency, 31
  connected in series, 87
  connection tube, design of, 31
  diameter, 15

[Column]
of analytical HSLC, 31
in preparative HSLC, 31
heavily loaded, 17, 64, 118
of infinite diameter, 15, 58
linear capacity of, 18
length, 31
linear capacity of, 58
loading capacity of, 57
materials, comparison of, 31
open tube in LC, 31
overloading of, 57
packed irregularly, 10
parallel, 85
permeability, 9, 11
estimation of, 13
parameter, 20
relationship between particle diameter and, 9, 13, 14
regeneration of, 60
sequential, 85
switching, automatic, 60
systems, multiple, 85
Combination of HSLC:
and MS, 100
with other chromatographic techniques, 60
Comparison of:
conventionally coated and chemically bonded packing materials, 41
different packing materials in the separation of polynuclear aromatic hydrocarbons, 79
GC and HSLC, 98
gradient elution, temperature programming, flow programming and coupled columns, 43
HSLC and TLC, 75, 78, 79, 110
isocratic and gradient elution, coupled columns, flow and temperature programming, 78
normal and reverse phase, ion-exchange HSLC in the analysis of steroids, 114
[Comparison of]
pellicular and conventional ion exchangers, 105
Zipax and small particle porous packing materials, 35
Computer:
in peak identification, 21
simulation, 27
system, on line, for peak identification, 61
Computing integrator, for HSLC, 61
Control of flow rate, 29
Controlled Pore Glass (CPG), 74
Control system, paper tape, 61
Corasil, 31, 33-35, 38, 59, 69, 71, 74, 77, 80, 91, 95, 100, 108, 115, 119-122
Corning Porous Glass, 34
Coronene, 79
Corticosteroids, 117, 118
in ointments and creams, 118
Corticosterone, 115
Cortisol, 115
Cortisone, 115, 116
Cortisone alcohol, 119
Cough preparation, active ingredients in, 96
Coulometry, constant potential, 102
Coupled HSLC-MS system, 94
Creatinine, 87
Crude oil, analysis of, 112
Cryptopine, 70
p-Cumaric acid, isomers, 66
Cyasterone, 117
3,5-Cyclic adenosine monophosphate, 105
Cyclohexanone, trace amounts in cyclohexanone oxime, 123

D

DDD, 108
DDT, 37, 108, 109

DEAE-cellulose, 106
Degassing, 23
Dehydroacetic acid, 67
Dehydroascorbic acid, 119
Deoxycytidine, in urine, 105
Deoxyinosine, 105
Deoxyribonucleotides, 105
Detection:
  of amino acids, 46
  nucleic acid constituents, 46
  simultaneous, at two wave-
    lengths, 45
Detectors, 43-57
  atomic absorption, 57
  capacitance, 56
  Christiansen-effect, 48, 113
  classification of, 43, 44
  colorimetric, 46
  control of temperature of, 44
  coulometric, 56, 102
  destructive FID, 50
  disk FID, 50
  electrochemical, 56, 76
  electrolytic conductivity,
    55, 56
  flow-fluorimeter, miniature,
    52
  fluorescence, 52, 53, 94, 120
    scanning spectrometry, 52
  gas density balance, 56
  heat of adsorption, 50-52
    computer simulation at, 50
    dual microadsorption, 51
    with gradient elution, 27
    modification of, 51
  improvement of stability and
    sensitivity of, 46
  interferometer, 48
  light absorption, 45-47
  light scattering, 56
  liquid scintillation, 89
  moving wire FID, 49, 83, 100,
    115, 122
    modification of, 49
  parallel, 44, 45
  polarographic, 53, 54, 56
    carbon-impregnated silicone
      rubber membrane electrode,
      53
[Detectors]
    dropping mercury electrode,
      53
    potentiostatic pulse tech-
      nique of, 53
  primary parameters of, 44
  radioactivity, 54, 55, 72
    efficiencies of, 54
  radiometric, 102
  rapid-scanning spectropho-
    tometry, 47
  recording balance, 56
  reference cell, 45, 48, 50,
    52
  refractometer:
    with gradient elution, 27
    types of, 47
  in series, 44
  thermometric, 51, 102
  transport, 48-50
  UV:
    for preparative work, 46
    types of cells for, 45
    with RI, dual, 48
  vapor pressure, 56
Dexamethasone, 117
Di-n-alkyl phthalates, 121
Diaminophenyl-methane iso-
    mers, 75
Diazepam, 95
Diazinon, metabolites of, 110
3,4-Dibenzyloxybenzaldehyde, 95
1-(3,4-Dibenzyloxyphenyl)-2-
    nitro-trans-prop-1-ene,
    95
2,5-Dichloroaniline, 76
3,4-Dichloroaniline, 76
2,4-Dichlorophenoxyacetic
    acid, 109
  esters of, 67
Dicyanoheptamethylcobyrinate,
    59
N,N-diethylaniline, 74, 75
N,N-diethylazoaniline, 76
N,N-diethyl-m-toluamide, 76
Digitoxigenin, 117
  glycosides of, 117
Diglycine, 73
Dihydrodiazepam, 95

3,6-Dihydroxytropine, 71
Diketogluconic acids, 65
Dimethylamine, 99
N,N-dimethyl-p-aminobenzene-benzoyl esters, 66
N,N-dimethylazoaniline, 76
7,12-dimethylbenzo(a)anthracene, 77
N,N-dimethyl-2,2-diphenylacetamide, metabolites of (see also ^{14}C-Diphenamid), 90
Dimethylphthalate, 64
N-(2,4-dinitrophenyl)-diethylamine, 99
Dinonyl phthalate, 26
Dioctyldiphenylamine, 75
Diolefins, 77
^{14}C-Diphenamid, metabolites of (see also N,N-dimethyl-2,2-diphenylacetamide), 90
m-Diphenyl, 35
Diphenylamine, 75
Diphenylhydantoin, 92
  determination through its conversion into diphenylketone, 93
  simultaneous measurement of phenobarbital and, 93
  urinary metabolites of, 93
Diphenylphthalate, 121
Diphenylsulfonates, 82
Disaccharides, 83
Distribution (see also Partition), 1
  coefficient, 7
    dependence on interaction forces, 6
  isotherm, 4, 8
    linearity of, 18
Dithianone, 109
Diuron, 106
  high-sensitivity detection of, 107
Donor-acceptor complex, 38
L-Dopa, 76
  metabolites of, 88
Dopamine, 76
Dowex, 66, 67, 106
Drug purity profiles, 97
Drug-treated tumor cells, metabolites from, 96
Durapak, 36, 74, 75, 79, 80, 90
Dyfonate, 109

E

Ecdysone, 117
Efficiency:
  effect of temperature on, 80
  a test mixture for column, 64
Eluotropic series, 42
Elution bandwidth, contribution of injection to, 30
Emetine, 68
Endrin, 108
Engine oils, polynuclear hydrocarbons in used, 80
Ephedrine, 92
Epichlorohydrin-bisphenol A epoxy resin, 60
EPN, insecticide, 121
Ergocalciferol, 119
Ergocristine, 71
Ergotamine, 71
Ergothioneine, 87
Essential oil, 26
Estradiol, 115
  derivatives of, 117
  glucosiduronic acid, 115
Estriol, 115
Estrogens:
  equine, 115
  in pregnancy and nonpregnancy urines, 115
Estrone, 115, 116
N-ethylaniline, 74, 75
Ethylorange, 28
Ethyl vanillin, 97-99
Extracolumn effects, 18, 31

F

Fatty acids:
  benzylderivatives of, 100
  in pig liver extract, 100

Fenuron, 106
Ferric(III) ion, 102
Ferulic acid, 66
Flavor compounds, 99, 100
  in boiled beef, 59
  in citrus oils and rums, 99
Flow programming, 28, 29, 92, 114
  automatic, 60
Fluocinolone acetonide, 118
Fluocinolone acetonide acetate, 118
Fluocionide ointment preparations, 118
Fluoranthrene, 77, 78
Fluorescamine reagent, 72
Fluorescent derivatives of compounds, 52
Folic acid, 119
Folpet, 109
Food:
  additives in soft drinks, 97
  colors, 99
  preservatives, 97
3-Formylrifampin, impurities in, 95
Fraction collector:
  automatic, 58
  in preparative HSLC, 58
Fractonitril, 17, 35, 77, 111
Freon, 76
Fructose, 82, 83, 85
Fucose, 83, 85
Fumaric acid, 65, 67
Furocoumarin, 91
2-Furoylglycine, in urine, 68

G

Galactose, 82, 83, 85
Gel permeation chromatography, definition of, 111
General elution problem, 43
Glucose, 26, 82, 83, 85
  oligomers, 83
Glucuronides, in urine, 90
Glutamine, 71, 72
Glutaric acid, 88
Glycerophosphate, 122
Glycerophosphorylcholine, 122
Glycine, 73
Glycolipids, 122
Glycols:
  diethylene, 63
  monoethylene, 63
  polyethylene, 113
Glycosides:
  nitrophenyl derivatives of, 83
  phenyl derivatives of, 83
Gradient elution, 5, 6, 59
  apparatus, specially designed, 67
  automatic, 27, 60
  computerized, 85
  computer program for, 26
  definition of, 24
  devices, 25
    multifunctional, 26
  incremental method of, 26
  methods of generating, 29
  optimum, 25
  pH control problems in, 27
  pressure-forced system, 67
  at radioactivity detectors, 54
  under conditions of axial equilibrium, 25
Griseofulvin, 96
G salt, 67

H

Hashish, 94
Heptachlor, 108
Heptaglycine, 73
Heroin, 70, 94
Hexachlorophene, 110
Hippuric acid, 87
Histidine, 72
Homogentistic acid, 88
Homovanillic alcohol, 88
Hop bitter acids, contribution of the double-bond to retention of, 100
Hop resins, 100
Human urine, analysis of, 59, 87

4-Hydroxy acetanilide, sulfate ester of, 89
2-Hydroxy aniline, sulfate ester of, 89
4-Hydroxy aniline, sulfate ester of, 89
5-Hydroxy-2-benzimidazolecarbamate, 89
p-Hydroxy-benzoates, alkyl-substituted, 97
p-Hydroxy-benzoic acid, 99
Hydroxybenzoic acids, isomeric, 111
2-Hydroxy-butyric acid, 87
Hydrocortisone, 116
  acetate, 116
4-Hydroxy-2,6-dimethylnicotinic acid, 67
6-Hydroxy-dopamine, 76
5-Hydroxymethyl-2-furoic acid, in urine, 68
Hydroxynalidixic acid, in human plasma and urine, 88
4-Hydroxy-phenylpyruvic acid, 88
8-Hydroxyquinoline, 69
5-Hydroxy-tryptamine, 76
5-Hydroxy-tryptophan, 76
Hyosciamine, 71
Hypoxanthine, 87, 105

I

Imidan, 109
Indoles, 87
  in biological fluids, 85
Insect juvenile hormone, 108
Interactions, between solute and stationary phase, 42
Interfacial tension, 6
Iodide, detection limit of, 56
Iodobenzenes, 80
Isopropyl carbanilate, metabolites (see also Propham), 89
4-Isopropylphenol, 110
Isocitric acid, 67
Isocyanate, oligomers, 113
Isohumulones, in beer, 100
Isoquinoline, 69, 75

K

Ketazolam, 95
α-Ketoglutaric acid, 67
  in urine, 68
17-Ketosteroids, DNPH-derivatives of, from urine and plasma, 117

L

Lactic acid, 67, 87
  in urine, 68
Lactose, 83, 85
  in milk, 99
Lannate methomyl, active ingredient of, 108
Lanthanides, 101
Lead(II) ion, 102
Leucine, 72
Lindane, 108
Linear capacity maximum, in LSC, 42
Linuron, 106
Lipids:
  optimization of the elution parameters, 122
  of rat blood cell, 122
LSD, 94

M

Maleic acid, 65
Malonic acid, 88
Maltose, 83
Mandelic acid, 88
Mannose, 82, 83
Mass spectrometry, 89
  coupling with HSLC, 57
Mass transfer:
  lateral, 31
  resistance constant, 13
  in stationary phase, 35
  between stationary and mobile phases, 35
Maximum sample resolution per unit time, 25

Meperidine, 95
Metal carbonyl complexes, 101
Metal chelation, 37
Metal-β-diketonates, 101
Metal ions, 102
Metallocarbones, 102
Methadone, 70
Methaqualone, 95
2-Methoxy-4,6-bis(isopropyl-amine)-S-triazine, urinary metabolites of (see also Prometone), 89
Methoxychlor, 108
1-Methoxynaphthalene, 78
2-Methoxynaphthalene, 78
2-Methylalanine, 75
Methyl-2-benzimidazolecarbamate, 89
Methyl benzoate, 65
N-Methyl carbamates, dansyl derivatives of, 107
p,p'-Methylenedianiline, 74
Methyl-4-hydroxy-2-benzimidazolecarbamate, 89
Methylmaleic acid, 67
Methylmalonic acid, 87
Methylorange, 28
Methyl palmitate, 26
Methyl-Parathion, 110
Methyl prednisolone, 117
Methyl prednisolone-21-acetate, 117
Methyl testosterone, 116
Methyl vanillin, 99
Metolazone, in urine, 90
Microporous silica gel (see Small-particle silica gel)
Minimum time analysis, 19, 20
Mobile phase:
  choice of, 6
  degassing of, 43
  presaturation of, 36, 43
  selection of, 42
    general criteria for, 42
    in preparative HSLC, 42
  water content in, 35, 42
Mononitrophenols, 110
Monosaccharides, 82, 83
Monuron, 106
Morphine, 70

N

Nalidixic acid, in human plasma and urine, 88
Naphazoline, 96
Naphthalene, 25, 77, 78
1-Naphthalene amine, 75
2-Naphthalene amine, 75
Naphthalene sulfonic acids, 68
  dye intermediate, 67
1-Naphtol, 108
Naphtols, 111
Narcotics, 95
Narcotine, 70
Niacin, 119
Nicotinamide adenine dinucleotide, impurities in, 105
Nitrazepam, 95
Nitroanilines, 74, 75
p-Nitrobenzaldehyde, 65
Nitrobenzene, 65
Nitroltriacetic acid, 98
1-Nitronaphthalene, 78
p-Nitrophenol, 110
N-Nitrosamine, 99
N-Nitroso-diethylamine, 99
o-Nitrotoluene, 78
p-Nitrotoluene, 37
Nonylphenolethyleneoxide, adducts, 113
Normal phase LLC, 39
19-Nortestosterone, 116
Nucleic acid constituents, 102-106
  optimum separation of, 103
Nucleotides:
  in cell, 91, 105
  modified, 105
  RNA terminal residues, 61
  in tissue, 105
Nylon 6, molecular weight distribution of, 113

O

6-(O-acetyl)-morphine, 71

O-aminobenzoic acid, 67
Octyl-2-naphthylamine, 75
Oils, edible, 100
Olefins, 77
Oligonucleotides, 106
Oligosaccharides, 83
Oligostyrenes, 31
3-O-methyl-D-glucose, 82
Optimization:
  general equation for, 19, 20
  practical approach to, 21
  procedures, 19
  ways of, 19
Optimum separation, in HSLC, 18
Optimum solvent program, design of, 25
Orcinol colorimetric method, 82, 83
Organic acids:
  in blood cells and biological fluids, 88
  in foods, 98
  serum, 87
  in wines and fruit juices, 98
Ornithine, 71
Orthophosphate, 122
Oxazepam, 95
Oxymethazoline, 96
Oxypurines, human plasma levels of the endogenous, 105

P

Packing materials:
  chemically bonded, 33, 35, 36
    brush-type, 15
    8-hydroxyquinoline, 37
    polar, 38
  comparison of, 34
  limitations of conventional, 36
  pellicular, 33, 46
    coated dynamically, 70
  reverse-phase, 37, 80
  in situ-formed polyurethane as, 39
  small-particle size porous, 34
Packing methods:
  balanced density slurry, 32
  comparison of, 33
  dynamic, 32
Papaverine, 70
Paraffins, 77
Parathion, 109, 110
Park-Johnson method, 83
Particle distribution, 16
Particle sizing:
  by flotation, 36
  by particle classifier, 36
Partition (see also Distribution), 1
  coefficient, 2, 3, 6
    dynamic determination of, 4
    prediction of, 4
    of steroids (see also Steroids), 114
  isotherm, linear, 2
  in LLC, 39
  ratio (see also Capacity factor), 3, 5
Peak capacity, 6
Pellicular packingmaterials:
  capacity of, 34
  ion exchange, 33
Pellidon, 33, 38, 75, 94, 97, 110
Pellionex, 72
Pellisieve, 38, 64, 65
Pellosil, 33, 64
Pellumina, 75, 119
Peptides, separation of, 73
Perisorb, 33-35
Permaphase, 26, 35, 64, 76, 80, 89, 94, 96, 107, 108, 115, 116, 118
Pesticides, 53, 106-110
  chlorinated hydrocarbon, 108
  partition coefficients of, 4
  residue analysis, 109
Phase ratio, 3
Phenacetin, 91, 92
  metabolic pathway of, evaluation of, 85
Phenanthrene, 37, 77, 78

Phenethylamines, 74, 94
Phenetidine, isomers, 75
Phenobarbital, 92
Phenolic acids, 66
Phenols, 31, 64, 110, 111
  effect of substituents on retention of, 110
  methyl-substituted, 111
Phenol-sulfuric acid method, 85
Phenoxyacetic acid-base herbicides, 108
Phentolamine, 96
Phenylacetic acids, OH-substituted, 88
Phenylalanine, 71, 72
Phenylbutazone, 97
Phenylene diamine, isomers, 76
Phenyl-D-galactopyranosides, 83
Phenylglycols, 88
N-Phenyl-2-naphthylamine, 75
6-Phenylphenol, 97
2-Phenyl-2-propanol, 64
Phosphatidylcholines, 122
Phospholipids, 122
Phosphorylcholine, 122
Phthalate plasticizers, 121
Phthalic acid, isomers, 121
  effect of pH on retention and resolution of, 121
Physiological fluids, fluorescent compounds in, 52
Picolines, isomeric, 69
Pig liver, analysis of, 99
Plant organic acids, 67
Plant sapogenins, 26
Plate height, 8, 11
  effect of fluid velocity on, 10
  effect of sample concentration on, in infinite diameter column, 17
  effect of sample size on, 17
    in infinite diameter column, 17
  reduced, 10, 20
    as function of reduced fluid velocity, 14, 15
    relation between column length and, 15
[Plate height]
    relation between particle diameter and, 12
Polyamines, in urine, 88
Polycar, 66
Poly-G 300 stationary phase, 69, 70
Polynuclear azaheterocyclic compounds, 69
Polysodium silicate, 113
Polystyrene, 60
Polythionates, in Wackenroder's solution, 122
Polyvinyl alcohols, 111
Ponasterone-A, 117
Poragel, 99
Porapak, 76, 78
Porasil, 34, 38, 59, 77
Pore size distribution, of pellicular packing materials, 36
Pork luncheon meat, analysis of, 99
Potassium sorbate, 97
Precision, in HSLC, 61
Prednisolone, 117
4-Pregnene-20β-ol-3-one, 114
Premium gasoline, separation of, 77
Preparative HSLC, 57-60, 91
  purity in, 59
  recycle technique in, 59, 60
Procaine, 95
Progesterone, 58, 59, 114, 116
Progestins, 114
Proline, 72
Prometone, urinary metabolites of (see also 2-Methoxy-4,6-bis)isopropylamino-(-S-triazine), 89
Propham, metabolites of (see also Isopropyl carbanilate), 89
Propionic acid, 87
Prostaglandins, 96
  in rat kidney, 96
Pseudotropine, 71
Pseudouridine, 87
Pulse damper, 29

Pulse damping, 44
Pulse-free pumping, 28
Pumps, 28
  diaphragm, 28
  multihead, 29
  pneumatic, 28
  reciprocating piston, 28
  screw-driven, 28
Purine, metabolism of, 87
Putrescine, 88
Pyrazalone, 99
Pyrene, 78
Pyrethins, 108
Pyrethrum insecticide extract, 58
Pyridine, 69
  isomers, substituted, 69
Pyridinol, 67
Pyridone, 67
Pyrimidine, metabolism of, 87
Pyruvic acid, 67, 87
  in urine, 68

Q

m-Quaterphenyl, 35, 78
o-Quaterphenyl, 78
Quinine, 70
Quinoline, 69, 75
  mixtures, hydrogenated, 68
m-Quinquephenyl, 35

R

Radioactive elements, separation of, 101
Radioactivity, conventional measurement methods in HSLC, 55
Rare earths, 101
Recycle technique, 96
Reservoirs, in preparative work, 23
Resins, phenol formaldehyde, 113
Resolution, 42
  in column of infinite diameter, 16
[Resolution]
  dependence on pressure, 10, 18
  effect of solvent gradient, particle size and cross-linkage on, 101
  general equation for, 20
  improvement of, 5, 25
  influence of wall effects on, 18
  maximum, for a given pressure, 10
  of pyridine isomers, 69
Retention, relative, 5, 14, 37
  contribution of factors to, 6
  dependence on ion exchange reactions, 18
Retention time, in minimum time analysis, 19
Retention volume, 3
  net, 3
    effect of sample size on, 18
  prediction of, 4
  sample equivalent, 17
  specific, 3, 4
Retinol, 119
Reverse-phase LLC, 39
Reviews on:
  application of HSLC in brewing, 100
  chemically bonded packing materials, 39
  chromatographic and biological aspects of phthalate esters, 121
  clinical use of HSLC, 91
  computation in chromatography, 61
  detectors in HSLC, 43
  HSLC, 23
  HSLC columns, 33
  HSLC equipments, 23
  pumps, 29
  routine analyses by HSLC, 63
  sample loading, 30
Rhamnose, 82, 83
Riboflavine, 99

Ribose, 83
R salt, 67

S

Salicylamide, 91
Sample:
  injection, 29
    stop flow, 30
    under pressure, 30
  loading, automatic, 60
  solubility, improvement of, 43
  valves, 30
Scaling up from analytical to preparative HSLC, 58
Schaffer's salt, 67
Scintillation, 55
Scopolamine, 71, 95
Sedatives, 95
Selectivity, 1
  in LSC, 7
  in the separation of alcohols, 64
  thermodynamic basis of, 6
Sensitivity limit of aliphatic aldehydes, 65
Sephadex, 113
  hydroxyalkoxypropyl, 120
Serine, 72
Serum uric acid, 89
Small molecules by HSGPC, 113
Small-particle silica gel, 64, 66, 72-74, 93, 94, 97-100, 110, 113, 115, 117, 119-121
  coated in situ, 76
Sodium benzoate, 97
Sodium naphthalene-2-sulfonate, 68
Sodium saccharide, 97
Solubility:
  parameter, 43
  of sample, 58
Solvent pairs, most widely used, 39, 40
Solvent strength parameter, 43
Sorbic acid, 97
Sorbose, 85
Spermidine, 88
  monoacetyl, 88
Spermine, 88
Spherosil, 59, 78, 79, 115
Spray, analysis of, 76
Squalane, stationary phase, 26, 77
Stationary phase:
  choice of, 6, 39
  diamond as, 79
  liquid mixtures as, 41
  loading, 16
    optimum level of, 39
    suggested, 39
  mass transfer in, 17
Steroids, 26, 113-118
  androgenic, 117
  in biological fluids, 61
  in body fluids, 118
  chromatography on CN-terminated stationary phases, 118
  corticoid, 114
    adrenal, 114, 115
    from plasma, 115
    trace amounts in blood and pregnancy urine, 114
  2,4-dinitrophenylhydrazine derivatives of, 114
  estrogenic:
    equine, 117
    trace amounts in blood and pregnancy urine, 114
  fermentation mixtures of, 118
  human, 117
  male sexual, 118
  partition coefficients of (see also Partition coefficients), 4, 114
  in rat serum, 115
  reverse-phase chromatography of, 113
Strychnine, 69
Succinic acid, 67
Sucrose, 82, 83, 85
Sugar-borate complexes, 85
Sulfadiazine, 94
Sulfaisomidine, 94

Sulfamerazine, 94
Sulfamethazine, 94
Sulfamethazone, 94
Sulfanilamide, 94
Sulfanilic acids, 75, 99
Sulfapyridine, 94
Sulfathiazine, 94
Sulfonylurea antidiabetic
agents, 95
Surfactants, 112
Surface-etched glass beads, 39

T

Taurine, 71, 72
Teloidine, 71
Temperature programming, 25, 90
automatic, 60
Terephthalic acid, 67
m-Terphenyl, 35
Testosterone, derivatives of,
117
Tetracycline antibiotics, 96
Tetrahydrozoline, 96
Tetralin, 78
Tetryl, 122
Thebaine, 70
Theobromine, 69
Theophilline, 69, 92
Thiamine, 99
1-Thioglycosides, 83
Thiohydroxamates, 107
Thioridazine, distribution
coefficients of, 96
Threourine, 72
Thydroidal iodoamino acids, 73
Time normalization, 19, 20
Time-share computer system, 61
Tocopherol (see also Vitamin
E), 118, 120
acetate, 119
Toluene, 31, 64, 78
Toluic acid, 67
m-Toluidine, 75
o-Toluidine, 75
Total aromatics present in mix-
tures, determination of,
79
Transcolumn inequilibrium, 16
Transfer of TLC separations to
HSLSC, 18
Tricarboxylic acid cycle
intermediates, 67
Triethyleneglycol stationary
phase, 76
Triglycerides, 100
Trimethyleneglycol stationary
phase, 64, 69, 74
2,4,6-Trinitrotoluene, 122
Tris(acetylacetonato)chromium
(III), 102
Tris(acetylacetonato)cobalt
(III), 102
Tristearine, 26
Trisulfapyrimidines, 95
Triton X45, 27
Tropanol, 34
Tropine, 71
Tryptophan, 71, 72
Tyrosine, 71, 72
metabolites of, 88

U

UCON 50HB55, 27
Uracil, 87
Urea, 71, 72
Uric acid, 87
Uridine, 87
Urinary polyamines, 76, 88
Urine, analysis of, 27, 61, 85
of normal newborns and
children, 87

V

Vanilla, 100
Vanillic acid, 88
Vanillin, 97, 98
Violaxanthin, 59
isomers of, 99
Vitamins, 118-120
fat-soluble, 118, 119
mineral tablet, 119
water-soluble, 118, 119
Vydac, 33, 64, 95

W

Wall effects, 18
Water environmental study,
HSLC in, 122

X

Xanthine, 87, 105
derivatives of, 92
Xylene, 61
Xylenol, isomers, 110
Xylidine, isomers, 74
Xylomethazoline, 96
Xylose, 82, 83, 85

Y

Yellow, 5, 99

Z

Zeocarb, 90
Zipax, 7, 13-15, 33-35, 64-68,
72, 76, 95, 96, 102,
108, 121
particle size distribution
for, 36
Zorbax, 58